Newnes Guide to Digital TV

Newnes Guide to Digital TV

Second edition

Richard Brice

Newnes

OXFORD AMSTERDAM BOSTON LONDON NEW YORK PARIS
SAN DIEGO SAN FRANSISCO SINGAPORE SYDNEY TOKYO

Newnes
An imprint of Elsevier Science
Linacre House, Jordan Hill, Oxford OX2 8DP
200 Wheeler Road, Burlington, MA 01803

First published 2000
Second edition 2003

British Library Cataloguing in Publication Data
A catalogue record for this book is available from the British Library

ISBN 0 7506 5721 9

For information on all Newnes publications, visit
our website at www.newnespress.com

Data manipulation by David Gregson Associates, Beccles, Suffolk
Printed and bound in Great Britain by Biddles Ltd, *www.biddles.co.uk*

Contents

Preface to the second edition

In the four years or so since I started work on the first edition of this book, digital television has changed from being a reality to a common place. My own family is proof of this. Whether the children are watching a wide choice of programmes via digital cable in Paris or, more recently, a similar choice carried by digital satellite in the UK, they have come to expect the benefits that digital television brings; more channels, better quality, interactivity etc. In addition to this ubiquity, there have been some real technical developments too. The inclusion of a hard drive within the set-top box, speculated on in the first edition, has now become a reality with the personal video recorder or PVR permitting the tapeless time-shifting of programmes and true video-on-demand (VOD) movie channels.

But it is not simply in the broadcast areas of television that we have seen huge advances in the last few years. The adoption of the DV camcorder and the proliferation of digital editing software on personal computer platforms has spread digital technology to the videographer as well as the broadcaster. Most of all, the DVD and the availability of reasonable priced wide-screen televisions have changed the public's perception of the quality boundaries of the experience of watching television. This edition, therefore, has sought to cover in more detail these trends and developments and, to that end, you will find herein expanded sections on DVD, the inclusion of sections on DV video compression and how it differs from MPEG.

As a general rule, nothing dates so comprehensively in a technical book as a chapter titled 'The future'! However, lest one believe that the pace of change makes it impossible for the non-specialist to follow, I am pleased to say that digital television has undergone several years of consolidation rather than revolution and the final chapter remains as relevant as it was when it was originally written.

Acknowledgements

Once again, I should like to thank those who are mentioned in the Preface to the First Edition, and add my thanks to Neil Sharpe of Miranda Technologies Ltd for permission to use the photographs of the Miranda DV-Bridge and Presmaster mixer. My family too, deserve to be thanked again for their forbearance in having a father who writes books instead of spending more time with them.

Richard Brice
Great Coxwell 2002

Preface to the first edition

Newnes Guide to Digital Television is written for those who are faced with the need to comprehend the novel world of digital television technology. Not since the 1960s – and the advent of colour television in Europe – have managers, technicians and engineers had to learn so much, so quickly; whether they work in the development laboratory, the studio or in the repair-shop. This book aims to cover the important principles that lie at the heart of the new digital TV services. I have tried to convey the broad architecture of the various systems and how these offer the functionalities they do. By concentrating on important principles, rather than presenting reams of detail, I hope the important ideas presented will 'stick in the mind' more obstinately than if I had adopted the opposite approach. I am also aware that there exists a new generation of engineers 'in the wings', as it were, to whom the world of digital television will be the only television they will know. For them, I have included a chapter on the important foundations of television as they have evolved in the first 60 years of this interesting and world-changing technology.

Acknowledgements

I think, if we're honest, most engineers who work in television would agree that the reality of transmitted digital television has crept up on us all. Like a lion, it has circled for the last 20 years, but the final pounce has taken many by surprise. In truth, I should probably have started writing *Newnes Guide to Digital Television* earlier than I did. But I don't think it's just retrospective justification to say that I hesitated because, even a short time ago, many technologies that are today's reality were still in the laboratory and research information was very thin indeed. There has therefore been the need to make up for lost time in the publishing phase. I should also like to thank Andy Thorne and Chris Middleton of Design Sphere of Fareham in the UK, who were extremely helpful and patient as I wrestled to understand current set-top box technology; their help was invaluable and greatly appreciated. It is with their permission that the circuit diagrams of the digital set-top receiver appear in Chapter 11.

Finally, my thanks and apologies to Claire, who has put up with a new husband cloistered in his study when he should have been repairing the bathroom door!

Richard Brice
Paris, 1999

1
Introduction

Digital television

Digital television is finally here ... today! In fact, as the DVB organization puts it on their web site, put up a satellite dish in any of the world's major cities and you will receive a digital TV (DTV) signal. Sixty years after the introduction of analogue television, and 30 years after the introduction of colour, television is finally undergoing the long-predicted transformation – from analogue to digital. But what exactly does digital television mean to you and me? What will viewers expect? What will we need to know as technicians and engineers in this new digital world? This book aims to answer these questions. For how and where, refer to Figure 1.1 – the *Newnes Guide to Digital Television* route map. But firstly, why digital?

Why digital?

I like to think of the gradual replacement of analogue systems with digital alternatives as a slice of ancient history repeating itself. When the ancient Greeks – under Alexander the Great – took control of Egypt, the Greek language replaced Ancient Egyptian and the knowledge of how to write and read hieroglyphs was gradually lost. Only in 1799 – after a period of 2000 years – was the key to deciphering this ancient written language found following the discovery of the Rosetta stone. Why was this knowledge lost? Probably because Greek writing was based on a written alphabet – a limited number of symbols doing duty for a whole language. Far better, then, than the seven hundred representational signs of Ancient Egyptian writing. Any analogue system is a representational system – a wavy current represents a wavy sound pressure and so on. Hieroglyphic electronics if you like! The handling and processing of continuous time-variable signals (like audio and video waveforms) in digital form has all the advantages of a precise symbolic code (an alphabet) over an older

1

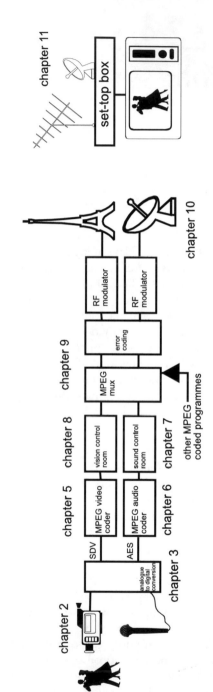

Figure 1.1 *Guide to Digital Television – Route map*

approximate representational code (hieroglyphs). This is because, once represented by a limited number of abstract symbols, a previously undefended signal may be protected by sending special codes, so that the digital decoder can work out when errors have occurred. For example, if an analogue television signal is contaminated by impulsive interference from a motorcar ignition, the impulses (in the form of white and black dots) will appear on the screen. This is inevitable, because the analogue television receiver cannot 'know' what is wanted modulation and what is not. A digital television can sort the impulsive interference from wanted signal. As television consumers, we therefore expect our digital televisions to produce better, sharper and less noisy pictures than we have come to expect from analogue models. (Basic digital concepts and techniques are discussed in Chapter 3; digital signal processing is covered in Chapter 4.)

More channels

So far, so good, but until very recently there was a down side to digital audio and video signals, and this was the considerably greater capacity, or bandwidth, demanded by digital storage and transmission systems compared with their analogue counterparts. This led to widespread pessimism during the 1980s about the possibility of delivering digital television to the home, and the consequent development of advanced analogue television systems such as MAC and PALplus. However, the disadvantage of greater bandwidth demands has been overcome by enormous advances in data compression techniques, which make better use of smaller bandwidths. In a very short period of time these techniques have rendered analogue television obsolescent. It's no exaggeration to say that the technology that underpins digital television is data compression or source coding techniques, which is why this features heavily in the pages that follow. An understanding of these techniques is absolutely crucial for anyone technical working in television today. Incredibly, data compression techniques have become so good that it's now possible to put many digital channels in the bandwidth occupied by one analogue channel; good news for viewers, engineers and technicians alike as more opportunities arise within and without our industry.

Wide-screen pictures

The original aspect ratio (the ratio of picture width to height) for the motion picture industry was 4:3. According to historical accounts, this shape was decided somewhat arbitrarily by Thomas Edison while working with George Eastman on the first motion picture film stocks. The 4:3 shape they worked with became the standard as the motion picture business grew. Today, it is referred to as the 'Academy Standard' aspect ratio. When the first experiments with broadcast television occurred in the 1930s, the $4:3$ ratio was used because of historical precedent. In cinema, $4:3$ formatted images persisted until the early 1950s, at which point Hollywood studios began to release 'wide-screen' movies. Today, the two most prevalent film formats are $1.85:1$ and $2.35:1$. The latter is sometimes referred to as 'Cinemascope' or 'Scope'. This presents a problem when viewing wide-screen cinema releases on a $4:3$ television. In the UK and in America, a technique known as 'pan and scan' is used, which involves cropping the picture. The alternative, known as 'letter-boxing', presents the full cinema picture with black bands across the top and bottom of the screen. Digital television systems all provide for a wide-screen format, in order to make viewing film releases (and certain programmes – especially sport) more enjoyable. Note that digital television services don't have to be wide-screen, only that the standards allow for that option. Television has decided on an intermediate wide-screen format known as $16:9$ ($1.78:1$) aspect ratio. Figure 1.2 illustrates the various film and TV formats displayed on $4:3$ and $16:9$ TV sets. Broadcasters are expected to produce more and more digital $16:9$ programming. Issues affecting studio technicians and engineers are covered in Chapters 7 and 8.

'Cinema' sound

To complement the wide-screen cinema experience, digital television also delivers 'cinema sound'; involving, surrounding and bone-rattling! For so long the 'Cinderella' of television, and confined to a 5-cm loudspeaker at the rear of the TV cabinet, sound quality has now become one of the strongest selling points of a modern television. Oddly, it is in the sound coding domain (and not the picture coding) that the largest differences lie between the European digital system and the American incarnation. The European DVB project opted to utilize the MPEG sound coding method, whereas the American infrastructure uses the AC-3 system due to Dolby Laboratories. For completeness, both of these are described in the chapters that follow; you will see that they possess many more similarities than differences: Each provides for multi-channel sound and for asso-ciated sound services; like simultaneous dialogue in alternate languages. But more channels mean more bandwidth, and that implies compression will be necessary in order not to overload our delivery medium. This is indeed the case, and audio compression techniques (for both MPEG and AC-3) are fully discussed in Chapter 6.

Associated services

Digital television is designed for a twenty-first century view of entertain-ment; a multi-channel, multi-delivery mode, a multimedia experience.

4:3 displayed 4:3

4:3 displayed 16:9

16:9 displayed 4:3

16:9 displayed 16:9

movie displayed 4:3

movie displayed 16:9

"scope" movie displayed 4:3

"scope" movie displayed 16:9

Figure 1.2 *Different aspect ratios displayed 4 : 3 and 16 : 9*

Such a complex environment means not only will viewers need help navigating between channels, but the equipment itself will also require data on what sort of service it must deliver: In the DTV standards, user-definable fields in the MPEG-II bitstream are used to deliver service information (SI) to the receiver. This information is used by the receiver to adjust its internal configuration to suit the received service, and can also be used by the broadcaster or service provider as the basis of an electronic programme guide (EPG) – a sort of electronic *Radio Times*! There is no limit to the sophistication of an EPG in the DVB standards; many broadcasters propose sending this information in the form of HTML pages to be parsed by an HTML browser incorporated in the set-top box. Both the structure of the MPEG multiplex and the incorporation of different types of data are covered extensively in Chapter 9.

This 'convergence' between different digital media is great, but it requires some degree of standardization of both signals and the interfaces between different systems. This issue is addressed in the DTV world as the degree of 'interoperability' that a DTV signal possesses as it makes the 'hops' from one medium to another. These hops must not cause delays or loss of picture and sound quality, as discussed in Chapter 10.

Conditional access

Clearly, someone has to pay for all this technology! True to their birth in the centralist milieu of the 1930s, vast, monolithic public analogue television services were nurtured in an environment of nationally instituted levies or taxes; a model that cannot hope to continue in the eclectic, diversified, channel-zapping, competitive world of today. For this reason, all DTV systems include mechanisms for 'conditional access', which is seen as vital to the healthy growth of digital TV. These issues too are covered in the pages that follow.

Transmission techniques

Sadly, perhaps, just as national boundaries produced differing analogue systems, not all digital television signals are exactly alike. All current and proposed DTV systems use the global MPEG-II standard for image coding; however, not only is the sound-coding different, as we have seen, but the RF modulation techniques are different as well, as we shall see in detail in later chapters.

Receiver technology

One phenomenon alone is making digital TV a success; not the politics, the studio or transmission technology, but the public who are buying the receivers – in incredible numbers! A survey of digital television would be woefully incomplete without a chapter devoted to receiver and set-top box technology as well as to digital versatile disc (DVD), which is ousting the long-treasured VHS machine and bringing digital films into increasing numbers of homes.

The future . . .

One experience is widespread in the engineering community associated with television in all its guises; that of being astonished by the rate of change within our industry in a very short period of time. Technology that has rcmained essentially the same for 30 years is suddenly obsolete, and a great many technicians and engineers are aware of being caught un-prepared for the changes they see around them. I hope that this book will help you feel more prepared to meet the challenges of today's television. But here's a warning; the technology's not going to slow down! Today's television is just that – for today. The television of next year will be different. For this reason I've included the last chapter, which outlines some of the current developments in MPEG coding that will set the agenda of television in the future. In this way I hope this book will serve you today and for some years to come.

2
Foundations of television

Of course, digital television didn't just spring fully formed from the ground! Instead it owes much to its analogue precursors. No one can doubt digital television represents a revolution in entertainment but, at a technological level, it is built on the foundations of analogue television. More than this, it inherited many presumptions and constraints from its analogue forebears. For this reason, an understanding of analogue television techniques is necessary to appreciate this new technology; hence the inclusion of this chapter. You will also find here a brief description of the psychological principles that underlie the development of this new television technology.

A brief history of television

The world's first fully electronic broadcast television service was launched by the BBC in London in November 1936. The system was the result of the very great work by Schocnberg and his team at the EMI Company. Initially the EMI system shared the limelight with Baird's mechanical system, but the latter offered poor quality by comparison and was quickly dropped in favour of the all-electronic system in February 1937. In the same year, France introduced a 455-line electronic system and Germany and Italy followed with a 441-line system. Oddly, the United States were some way behind Europe, the first public television service being inaugurated in New York in 1939; a 340-line system operating at 30 frames/second. Two years later, the United States adopted the (still current) 525-line standard.

Due to the difficulty of adequate power-supply decoupling, early television standards relied on locking picture rate to the AC mains frequency as this greatly ameliorated the visible effects of hum. Hence the standards schism that exists between systems of American and European origin (the AC mains frequency is 60 Hz in the North America rather than 50 Hz in Europe). In 1952 the German GERBER system was

8

proposed in order to try to offer some degree of harmonization between American and European practice. It was argued that this would ease the design of standards conversion equipment, and thereby promote the greater transatlantic exchange of television programmes; as well as enabling European television manufacturers the opportunity to exploit the more advanced American electronic components. To this end, the line frequency of the GERBER system was chosen to be very close to the 525-line American system but with a frame rate of 50, rather than 60, fields per second. The number of lines was thereby roughly defined by

$$(525 \times 60)/50 = 630$$

The GERBER system was very gradually adopted throughout Europe during the 1950s and 1960s.

The introduction of colour

Having been a little slow off the mark in launching an electronic TV service, television in the USA roared ahead with the introduction, in 1953, of the world's first commercial colour television service, in which colour information is encoded in a high-frequency subcarrier signal. Standardized by the National Television System Committee, this system is known world-wide as the NTSC system. Eight years later in France, Henri de France invented the Sequenticl Couleur a Memoire system (SECAM) which uses two alternate subcarriers and a delay-line 'memory store'. Although SECAM requires a very greatly more complicated receiver than NTSC (a not inconsequential consideration in 1961), it has the advantage that the colour signal can suffer much greater distortion without perceptible consequences. At about the same time – and benefiting from the technology of ultrasonic delay lines developed for SECAM – Dr Walter Bruch invented the German PAL system, which is essentially a modified NTSC system. PAL retains some of the robustness of SECAM, whilst offering something approaching the colour fidelity of NTSC. Colour television was introduced in the UK, France and Germany in 1967, 14 years after the introduction of the NTSC system.

Now let's look at some of the perceptual and engineering principles which underlie television; analogue and digital.

The physics of light

When an electromotive force (EMF) causes an electric current to flow in a wire, the moving electric charges create a magnetic field around the wire. Correspondingly, a moving magnetic field is capable of creating an EMF in an electrical circuit. These EMFs are in the form of voltages constrained to dwell inside electric circuits. However, they are special cases of electric fields. The same observation could be phrased: a moving electric field creates a magnetic field and a moving magnetic field creates an electric field. This affords an insight into a form of energy that can propagate through a vacuum (i.e. without travelling in a medium) by means of an endless reciprocal exchange of energy shunted backwards and forwards between electric and magnetic fields. This is the sort of energy that light is. Because it is itself based on the movement of electricity, it's no surprise that this type of energy has to move at the same speed that electricity does – about 300 million metres per second. Nor that it should be dubbed electromagnetic energy, as it propagates or radiates through empty space in the form of reciprocally oscillating magnetic and electric fields known as electromagnetic waves.

Although the rate at which electromagnetic energy radiates through space never changes, the waves of this energy may vary the rate at which they exchange electric field for magnetic field and back again. Indeed, these cycles vary over an enormous range. Because the rate at which the energy moves is constant and very fast, it's pretty obvious that, if the rate of exchange is relatively slow, the distance travelled by the wave to complete a whole cycle is relatively vast. The distance over which a wave of electromagnetic energy completes one cycle of its repeating pattern of exchanging fields is known as the wavelength of the electromagnetic energy. It's fair to say that the range of wavelengths of electromagnetic energy boggle the human mind, for they stretch from cosmic rays with wavelengths of a hundred million millionth of a metre to the energy radiated by AC power lines with wavelengths of a million million metres! For various physical reasons only a relatively small region of this huge range of energy, which floods from all over the universe and especially from our sun, arrives on the surface of the earth. The range that does arrive has clearly played an important role in the evolution of life, since the tiny segment of the entire diapason is the range to which we are attuned and that we are able to make use of. A small part of this small range is the range we call light. Wavelengths longer than light, extending to about a millimetre, we experience as heat.

Physiology of the eye

The wavelengths the human eye perceives extend only from about 380 nm to about 780 nm, in frequency terms, just over one octave. Visual experience may occur by stimulation other than light waves – pressure on the eyeball, for example; an observation that indicates that the experience of light is a quality produced by the visual system. Put another way, there's nothing special about the range of electromagnetic wavelengths 380–780 nm, it's just that we experience them differently from all the others. We shall see that colour perception, too, is a function of the perceptual system and not a physical attribute of electromagnetic radiation.

Physiologically, the eye is often compared to a camera because they both consist of a chamber, open at one end to let in the light, and a variable lens assembly for focusing an image on a light-sensitive surface at the rear of the chamber. In the case of the camera, the light-sensitive material is film; in the case of the eye, the retina. Figure 2.1 illustrates the human eye in cross-section.

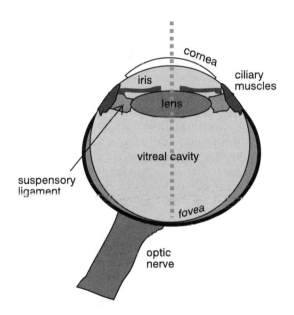

Figure 2.1   *The physiology of the eye*

A rough 2.5-cm sphere, the human eye bulges at the front in the region of the cornea – a tough membrane that is devoid of a blood supply in order to maintain good optical properties. The cornea acts with the lens within the eye to focus an image on the light-sensitive surface at the rear of the vitreal cavity. The eye can accommodate (or focus images at different distances) because the lens is not rigid but soft, and its shape may be modified by the action of the ciliary muscles. These act via the suspensory ligament to flatten the lens in order to view distant objects, and relax to view near objects. The iris, a circular membrane in front of the lens, is the pigmented part of the eye that we see from the outside. The iris is the eye's aperture control, and controls the amount of light entering the eye through the opening in the iris known as the pupil.

The light-sensitive surface at the back of the eye known as the retina has three main layers:

1. Rods and cones – which are photosensitive cells that convert light energy into neural signals;
2. Bipolar cells, which make synaptic connections with the rods and cones;
3. Ganglion cells, which form the optic nerve through which the visual signals are passed to the vision-processing regions of the brain.

Experiments with horseshoe crabs, which possess a visual system notably amenable to study, have revealed that the light intensity falling upon each visual receptor is conveyed to the brain by the rate of nerve firings. In our present context, the cells of interest are the 100 million cylindrical rods and the 6 million more bulbous cones that may be found in one single retina. The cones are only active in daylight vision, and permit us to see both achromatic colours (white, black and greys – known as luminance information) and colour. The rods function mainly in reduced illumination, and permit us to see only luminance information. So it is to the cones, which are largely concentrated in the central region of the retina known as the fovea, that we must look for the action of colour perception.

Psychology of vision – colour perception

Sir Isaac Newton discovered that sunlight passing through a prism breaks into the band of multicoloured light that we now call a spectrum. We perceive seven distinct bands in the spectrum:

red, orange, yellow, green, blue, indigo, violet

We see these bands distinctly because each represents a particular band of wavelengths. The objects we perceive as coloured are perceived thus because they too reflect a particular range of wavelengths. For instance, a daffodil looks yellow because it reflects predominantly wavelengths in the region 570 nm. We can experience wavelengths of different colour because the cones contain three photosensitive chemicals, each of which is sensitive in three broad areas of the light spectrum. It's easiest to think of this in terms of three separate but overlapping photochemical processes; a low-frequency (long wavelength) *red* process, a medium-frequency *green* process and a high-frequency *blue* process (as electronic engineers, you might prefer to think of this as three shallow-slope band-pass filters!). When light of a particular frequency falls on the retina, the action of the light reacts selectively with this frequency-discriminating mechanism. When we perceive a red object, we are experiencing a high level of activity in our long wavelength (low-frequency) process and low levels in the other two. A blue object stimulates the short wavelength or high-frequency process, and so on. When we perceive an object with an intermediate colour, say the yellow of the egg yoke, we experience a mixture of two chemical process caused by the overlapping nature of each of the frequency-selective mechanisms. In this case, the yellow light from the egg causes stimulation in both the long wavelength *red* process and the medium wavelength *green* process. Because human beings possess three separate colour vision processes, we are classified as trichromats. People afflicted with colour blindness usually lack one of the three chemical responses in the normal eye; they are known a dichromats, although a few rare individuals are true monochromats. What has not yet been discovered, amongst people or other animals, is a more-than-three colour perception system. This is lucky for the engineers who developed colour television!

Metamerism – the great colour swindle

The fact that our cones only contain three chemicals is the reason that we may be fooled into experiencing the whole gamut of colours with the combination of only three so-called primary colours. The television primaries of red, green and blue were chosen because each stimulates only one of the photosensitive chemicals found in the cone cells. The great colour television swindle (known technically at metamerism) is that we can, for instance, be duped into believing we are seeing yellow by activating both the red and green tube elements simultaneously – just as would a pure yellow source (see Figure 2.2). Similarly, we may be hoodwinked into seeing light-blue cyan with the simultaneous activation of green and blue. We can also be made to experience paradoxical colours like magenta by combining red and blue, a feat that no pure light

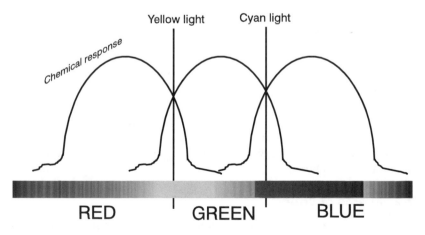

Figure 2.2   *Photochemical response of the chemicals in the eye*

source could ever do! This last fact demonstrates that our colour perception system effectively 'wraps-around', mapping the linear spectrum of electromagnetic frequencies into a colour circle, or a colour space. It is in this way that we usually view the science of colour perception; we can regard all visual sense as taking place within a colour three-space. A television studio vectorscope allows us to view colour three-space end-on, so it looks like a hexagon (Figure 2.3b). Note that each colour appears at a different angle, like the numbers on a clock face. 'Hue' is the term used in image processing and television to describe a colour's precise location on this locus. 'Saturation' is the term used to describe the amount a pure colour is diluted by white light. The dashed axis shown in Figure 2.3a is the axis of pure luminance. The more a particular shade moves towards this axis from a position on the boundary of the cube, the more a colour is said to be de-saturated.

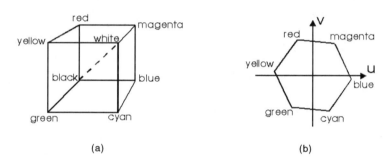

(a)   (b)

Figure 2.3   (a) *Colour three-space;* (b) *Colour three-space viewed end-on as in a TV vectorscope*

Persistence of vision

The human eye exhibits an important property that has great relevance to the film and video industries. This property is known as the persistence of vision. When an image is impressed upon the eye, an instantaneous cessation of the stimulus does not result in a similarly instantaneous cessation of signals within the optic nerve and visual processing centres. Instead, an exponential 'lag' takes place, with a relatively long time required for total decay. Cinema and television have exploited this effect for over 100 years.

The physics of sound

Sound waves are pressure variations in the physical atmosphere. These travel away at about 300 metres per second in the form of waves, which spread out like ripples on a pond. In their journey, these waves collide with the walls, chairs, tables – whatever – and make them move ever so slightly. The waves are thus turned into heat and 'disappear'. These waves can also cause the fragile membrane of the eardrum to move. Exactly what happens after that is a subject we'll look at later in the chapter. All that matters now is that this movement is experienced as the phenomenon we call hearing.

It is a demonstrable property of all sound sources that they oscillate: an oboe reed vibrates minutely back and forth when it is blown; the air inside a flute swells and compresses by an equal and opposite amount as it is played; a guitar string twangs back and forth. Each vibration is termed a cycle. The simplest sound is elicited when a tone-producing object vibrates backwards and forwards, exhibiting what physicists call simple harmonic motion. When an object vibrates in this way it follows the path traced out in Figure 2.4; known as a sine wave.

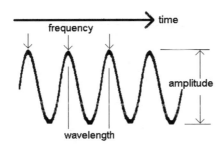

Figure 2.4 *Sine wave*

Such a pure tone, as illustrated, actually sounds rather dull and characterless. But we can still vary such a sound in two important ways. First, we can vary the number of cycles of oscillation that take place per second. Musicians refer to this variable as pitch; physicists call it frequency. The frequency variable is referred to in hertz (Hz), meaning the number of cycles that occur per second. Secondly, we can alter its loudness; this is related to the size, rather than the rapidity, of the oscillation. In broad principle, things that oscillate violently produce loud sounds. This variable is known as the amplitude of the wave.

Unfortunately, it would be pretty boring music that was made up solely of sine tones despite being able to vary their pitch and loudness. The waveform of a guitar sound is shown in Figure 2.5. As you can see, the guitar waveform has a fundamental periodicity like the sine wave, but much more is going on. If we were to play and record the waveform of other instruments each playing the same pitch note, we would notice a similar but different pattern; the periodicity would remain the same, but the extra small, superimposed movements would be different. The term

Figure 2.5 *Guitar waveform*

we use to describe the character of the sound is 'timbre', and the timbre of a sound relates to these extra movements which superimpose themselves upon the fundamental sinusoidal movement that determines the fundamental pitch of the musical note. Fortunately these extra movements are amenable to analysis too; in fact, in a quite remarkable way.

Fourier

In the eighteenth century, J.B. Fourier – son of a poor tailor who rose ultimately to scientific advisor to Napoleon – showed that any signal that can be generated can be alternatively expressed as a sum of sinusoids of various frequencies. With this deduction, he gave the world a whole new way of comprehending waveforms. Previously only comprehensible as a time-based phenomena, Fourier gave us new eyes to see with. Instead of thinking of waveforms in the time base (or the time domain) as we see them displayed on an oscilloscope, we may think of them in the frequency base (or the frequency domain) comprised of the sum of various sine waves of different amplitudes and phase.[1] In time, engineers have given us the tools to 'see' waveforms expressed in the frequency domain too. These are known as spectrum analysers or, eponymously, as Fourier analysers (see Figure 2.6). The subject of the Fourier transform, which bestows the ability to translate between these two modes of description, is so significant in many of the applications considered hereafter that a whole section is devoted to this transform in Chapter 4.

Transients

The way a musical note starts is of particular importance in our ability to recognize the instrument on which it is played. The more characteristic and sharply defined the beginning of a note, the more rapidly we are able to determine the instrument from which it is elicited. This bias toward transient information is even evident in spoken English, where we use about 16 long sounds (known as phonemes) against about 27 short

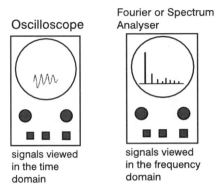

Figure 2.6   *A time-domain and frequency domain representation*

phonemes. Consider the transient information in a vocalized list of words that end the same way; coat, boat, dote, throat, note, wrote, tote and vote for instance! Importantly, transients too can be analysed in terms of a combination of sinusoids of differing amplitudes and phases using the Fourier integral as described above.

Physiology of the ear

Studies of the physiology of the ear reveal that the process of Fourier analysis, referred to earlier, is more than a mere mathematical conception. Anatomical and psychophysiological studies have revealed that the ear executes something very close to a mechanical Fourier analysis on the sounds it collects, and passes a frequency domain representation of those sounds on to higher neural centres. An illustration of the human ear is given in Figure 2.7.

After first interacting with the auricle or pinna, sound waves travel down the auditory canal to the eardrum. The position of the eardrum marks the boundary between the external ear and the middle ear. The middle ear is an air-filled cavity housing three tiny bones; the hammer, the anvil and the stirrup. These three bones communicate the vibrations of the eardrum to the oval window on the surface of the inner ear. Due to the manner in which these bones are pivoted, and because the base of the hammer is broader than the base of the stirrup, there exists a considerable mechanical advantage from eardrum to inner ear. A tube runs from the base of the middle ear to the throat; this is known as the Eustachian tube. Its action is to ensure that equal pressure exists on either side of the eardrum, and it is open when swallowing. The inner ear is formed in two sections; the cochlea (the spiral structure which looks like a snail's shell) and the three semicircular canals. These latter structures are involved with the sense of balance and motion.

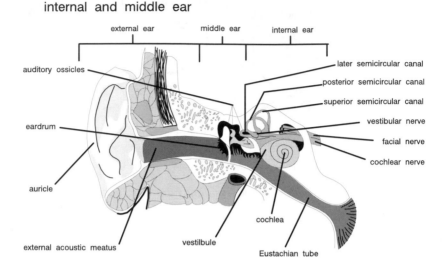

internal and middle ear

external ear | middle ear | internal ear

auditory ossicles
eardrum
auricle
external acoustic meatus
vestilbule

later semicircular canal
posterior semicircular canal
superior semicircular canal
vestibular nerve
facial nerve
cochlear nerve
cochlea
Eustachian tube

Figure 2.7   *The physiology of the ear*

The stirrup is firmly attached to the membrane that covers the oval window aperture of the cochlea. The cochlea is full of fluid and is divided along its entire length by the Reissner's membrane and the basilar membrane, upon which rests the organ of Corti. When the stirrup moves, it acts like a piston at the oval window and this sets the fluid within the cochlea into motion. This motion, trapped within the enclosed cochlea, creates a standing wave pattern – and therefore a distortion – in the basilar membrane. Importantly, the mechanical properties of the basilar membrane change considerably along its length. As a result, the position of the peak in the pattern of vibration varies depending on the frequency of stimulation. The cochlea and its components work thus as a frequency-to-position translation device. Where the basilar membrane is deflected most, there fire the hair cells of the organ of Corti; these interface the afferent neurones that carry signals to the higher levels of the auditory system. The signals leaving the ear are therefore in the form of a frequency domain representation. The intensity of each frequency range (the exact nature and extent of these ranges is considered later) is coded by means of a pulse rate modulation scheme.

Psychology of hearing

Psychoacoustics is the study of the psychology of hearing. Look at Table 2.1.

It tells us a remarkable story. We can hear, without damage, a ratio of sound intensities of about 140 dB or 1 : 1000 000 000 000. The quietest

Table 2.1

Phons (dB)	Noise sources
140	Gunshot at close range
120	Loud rock concert, jet aircraft taking off
100	Shouting at close range, very busy street
90	Busy city street
70	Average conversation
60	Typical small office or restaurant
50	Average living room, quiet conversation
40	Quiet living room, recording studio
30	Quiet house in country
20	Country area at night
0	Threshold of hearing

whisper we can hear is a billionth (10^{12}) of the intensity of the sound of a jet aircraft taking off heard at close range. In engineering terms, you could say human audition is equivalent to a true 20-bit system – 16 times better than the signal processing inside a compact disc player! Interestingly, the tiniest sound we can hear occurs when our eardrums move less than the diameter of a single atom of hydrogen. Any more sensitive, and we would be kept awake at night by the sound of the random movement of the nitrogen molecules within the air around us. In other words, the dynamic range of hearing is so wide as to be up against fundamental physical limitations.

Masking

The cochlea and its components work as a frequency-to-position translation device, the position of the peak in the pattern of vibration on the basilar membrane depending on the frequency of stimulation. Because of this it goes without saying that the position of this deflection cannot be vanishingly small – it has to have some dimension. This might lead us to expect that there must be a degree of uncertainty in pitch perception and indeed there is, although it's very small indeed, especially at low frequencies. This is because the afferent neurones, which carry signals to the higher levels of the auditory system, 'lock on' and fire together at a particular point in the deflection cycle (the peak). In other words, a phase-detection frequency discriminator is at work. This is a truly wonderful system, but it has one drawback; due to the phase-locking effect, louder signals will predominate over smaller ones, masking a quieter sound in the same frequency range. (Exactly the same thing happens in FM radio, where this phenomenon is known as capture effect.) The range of frequencies over which one sound can mask another is known as a critical band, a concept due to Fletcher (quoted in Moore, 1989). Masking is very familiar to us in our daily lives. For instance, it accounts for why we cannot hear someone whisper when someone else is shouting. The masking effect of a pure tone gives us a clearer idea about what's going on. Figure 2.8 illustrates the unusual curve that delineates the masking level in the presence of an 85 dBSPL tone. All sounds underneath the curve are effectively inaudible when the tone is present! Notice that a loud, pure sound only masks a quieter one when the louder sound is lower in frequency than the quieter, and only then when both signals are relatively close in frequency. Wideband sounds have a correspondingly wide masking effect. This too is illustrated in Figure 2.8, where you'll notice the lower curve indicates the room noise in dBSPL in relation to frequency for an average room-noise figure of 45 dBSPL. (Notice that the noise level is predominantly low frequency, a sign that the majority of the noise in modern life is mechanical in origin.) The nearly parallel line above this room-noise curve indicates the masking threshold. Essentially this illustrates the intensity level, in dBSPL, to which a tone of the indicated frequency would need to be raised in order to become audible. The phenomenon of masking is important to digital audio compression, as we shall see in Chapter 7.

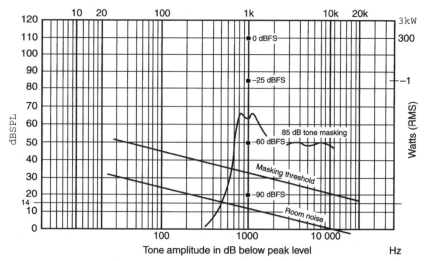

Figure 2.8   *Simultaneous masking effect of a single pure tone masking*

Temporal masking

Virtually all references in the engineering literature refer cheerfully to an effect known as temporal masking in which a sound of sufficient amplitude will mask sounds immediately preceding or following it in time; as illustrated in Figure 2.9. When sound is masked by a subsequent signal the phenomenon is known as backward masking, and typical quoted figures for masking are in the range of 5–50 ms. The masking effect that follows a sound is referred to as forward masking and may last as long 50–200 ms, depending on the level of the masker and the masked stimulus.

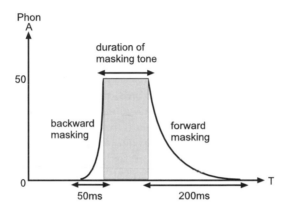

Figure 2.9   *The phenomenon of temporal masking*

Unfortunately, the real situation with temporal masking is more complicated, and a review of the psychological literature reveals that experiments to investigate backward masking in particular depend strongly on how much practice the subjects have received – with highly practised subjects showing little or no backward masking (Moore, 1989). Forward masking is, however, well defined (although the nature of the underlying process is still not understood), and can be substantial even with highly practised subjects.

Film and television

Due to the persistence of vision, if the eye is presented with a succession of still images at a sufficiently rapid rate, each frame differing only in the positions moving within a fixed frame of reference, the impression is gained of a moving image. In a film projector each still frame of film is drawn into position in front of an intense light source whilst the source of light is shut-off by means of a rotating shutter. Once the film frame has stabilized, the light is allowed through – by opening the shutter – and the image on the frame is projected upon a screen by way of an arrangement of lenses. Experiments soon established that a presentation rate of about 12 still frames per second was sufficiently rapid to give a good impression of continuously flowing movement, but interrupting the light source at this rate caused unbearable flicker. This flicker phenomenon was also discovered to be related to the level of illumination; the brighter the light being repetitively interrupted, the worse the flicker. Abetted by the low light output from early projectors, this led to the first film frame-rate standard of 16 frames per second (fps); a standard well above that required simply to give the impression of movement and sufficiently rapid to ensure flicker was reduced to a tolerable level when used with early projection lamps. As these lamps improved flicker became more of a problem, until an ingenious alteration to the projector fixed the problem. The solution involved a modification to the rotating shutter so that, once the film frame was drawn into position, the shutter opened, then closed, then opened again, before closing a second time for the next film frame to be drawn into position. In other words, the light interruption frequency was raised to twice that of the frame rate. When the film frame rate was eventually raised to the 24 fps standard that is still in force to this day, the light interruption frequency was raised to 48 times per second – a rate that enables high levels of illumination to be employed without causing flicker.

Television

To every engineer, the cathode ray tube (CRT) will be familiar enough from the oscilloscope. The evacuated glass envelope contains an electrode assembly and its terminations at its base, whose purpose it is to shoot a beam of electrons at the luminescent screen at the other end of the tube. This luminescent screen fluoresces to produce light whenever electrons hit it. In an oscilloscope, the deflection of this beam is effected by means of electric fields – a so-called electrostatic tube. In television, the electron beam (or beams in the case of colour) is deflected by means of magnetic fields caused by currents flowing in deflection coils wound around the neck of the tube where the base section meets the flare. Such a tube is known as an electromagnetic type.

Just like an oscilloscope, without any scanning currents the television tube produces a small spot of light in the middle of the screen. This spot of light can be made to move anywhere on the screen very quickly by the application of the appropriate current in the deflection coils. The brightness of the spot can be controlled with equal rapidity by altering the rate at which electrons are emitted from the cathode of the electron gun assembly. This is usually effectuated by controlling the potential between the grid and the cathode electrodes of the gun. Just as in an electron tube or valve, as the grid electrode is made more negative in relation to the cathode, the flow of electrons to the anode is decreased. In the case of the CRT, the anode is formed by a metal coating on the inside of the tube flare. A decrease in grid voltage – and thus anode current – results in a darkening of the spot of light. Correspondingly, an increase in grid voltage results in a brightening of the scanning spot.

In television, the bright spot is set up to move steadily across the screen from left to right (as seen from the front of the tube). When it has completed this journey it flies back very quickly to trace another path across the screen just below the previous trajectory. (The analogy with the movement of the eyes as they scan text during reading can't have escaped you!) If this process is made to happen sufficiently quickly, the eye's persistence of vision combined with an afterglow effect in the tube phosphor conspire to fool the eye, so that it does not perceive the moving spot but instead sees a set of parallel lines drawn on the screen. If the number of lines is increased, the eye ceases to see these as separate too – at least from a distance – and instead perceives an illuminated rectangle of light on the tube face. This is known as a raster. In the broadcast television system employed in Europe, this raster is scanned twice in a 25th of a second. One set of 312.5 lines is scanned in the first 1/50th of a second, and a second interlaced set – which are not superimposed but are staggered in the gaps in the preceding trace – is scanned in the second 1/50th of a second. The total number of lines is thus 625. In North America, a total of 525 lines (in two interlaced passes of 262.5) are scanned in 1/30th of a second. Figure 2.10 illustrates a simple 13-line interlaced display. The first field of 6.5 lines is shown in black, and the second field in grey. All line retraces ('flybacks') are shown as dashed lines. You can see that the two half lines exist at the start of the second field and the end of the first. Note the long flyback at the end of the second field.

This may seem like a complicated way of doing things, and the adoption of interlace has caused television engineers many problems over the years. Interlace was adopted in order to accomplish a two to one

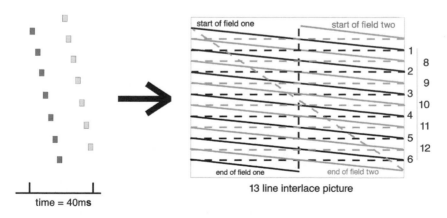

time = 40ms

Figure 2.10 *A simple interlaced 13-line display*

reduction in the bandwidth required for television pictures with very little noticeable loss of quality. It is thus a form of perceptual coding – what we would call today a data compression technique! Where bandwidth is not so important, as in computer displays, non-interlaced scanning is employed. Note also that interlace is, in some respects, the corollary of the double exposure system used in the cinema to raise the flicker frequency to double the frame-rate.

Television signals

The analogue television signal must do two things, the first is obvious, the second less so. First, it must control the instantaneous brightness of the spot on the face of the cathode ray tube in order that the brightness changes that constitute the information of the picture may be conveyed. Secondly, it must control the raster scanning so that the beam travels across the tube face in synchronism with the tube within the transmitting camera, otherwise information from the top left-hand side of the televised scene will not appear in the top left-hand side of the screen and so on! In the analogue television signal this distinction between picture information and scan synchronizing information (known as sync–pulse information) is divided by a voltage level known as black level. All information above black level relates to picture information; all information below relates to sync information. By this clever means, because all synchronizing information is below black level the electron beam therefore remains cut-off – and the screen remains dark – during the sync information. (In digital television the distinction between data relating to picture modulation and sync is established by a unique code word preamble that identifics the following byte as a sync byte, as we shall see.)

H sync and V sync

The analogy between the eye's movement across the page during reading and the movement of the scan spot in scanning a tube face has already been made. Of course, the scan spot doesn't move onto another page like the eyes do once they have reached the bottom of the page, but it does have to fly back to start all over again once it has completed one whole set of lines from the top to the bottom of the raster. The spot thus flies back in two possible ways; a horizontal retrace (between lines) and a vertical retrace (once it has completed one whole set of lines and is required to start all over again on another set). Obviously, to stay in synchronism with the transmitting camera, the television receiver must be instructed to perform both horizontal retrace and vertical retrace at the appropriate times – and furthermore, not to confuse one instruction for the other!

It is for this reason that there are two types of sync information, known reasonably enough as horizontal and vertical. Inside the television monitor these are treated separately, and respectively initiate and terminate the horizontal and vertical scan generator circuits. These circuits are, in principle, ramp or sawtooth generator circuits. As the current gradually increases in both the horizontal and vertical scan coils, the spot is made to move from left to right and top to bottom, the current in the top to bottom circuit growing 312.5 times more slowly than in the horizontal deflection coils so that 312.5 lines are drawn in the time it takes the vertical deflection circuit to draw the beam across the vertical extent of the tube face.

The complete television signal is illustrated in Figures 2.11 and 2.12, which display the signal using two different timebases. Notice the amplitude level, which distinguishes the watershed between picture information and sync information. Known as black level, this voltage is

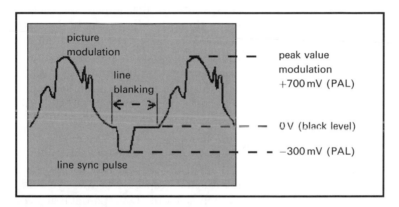

Figure 2.11   *Analogue TV signal viewed at line frequency*

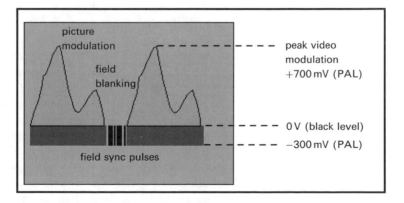

Figure 2.12   *Analogue TV signal viewed at frame frequency*

set to a standard 0 V. Peak white information is defined to not go beyond a level 0.7 V above this reference level. Sync information, the line or horizontal sync 4.7 microsecond pulse, is visible in the figure, and should extent 0.3 V below the black reference level. Note also that the picture information falls to the black level before and after the sync pulse. This interval is necessary because the electron beam cannot instantaneously retrace to the left-hand side of the screen to re-start another trace. It takes a little time – about 12 microseconds. This period, which includes the duration of the 4.7 microsecond line-sync pulse, during which time the beam current is controlled 'blacker-than-black', is known as the line-blanking period. A similar – much longer – period exists to allow the scan spot to return to the top of the screen once a whole vertical scan has been accomplished. This interval being known as the field-blanking or vertical interval.

Looking now at Figure 2.12, a whole 625 lines are shown, in two fields of 312.5 lines. Notice the wider sync pulses that appear between each field. In order that a monitor may distinguish between horizontal and vertical sync, the duration of the line-sync pulses is extended during the vertical interval (the gap in the picture information allowing for the field retrace) and a charge-pump circuit combined with a comparator is able to detect these longer pulses as different from the shorter line-sync pulses. This information is sent to the vertical scan generator to control the synchronism of the vertical scan.

Colour television

From the discussion of the trichromatic response of the eye and the discussion of the persistence of vision, it should be apparent that a colour scene may be rendered by the quick successive presentation of the red, green and blue components of a colour picture. Provided these images are displayed frequently enough, the impression of a full colour scene is indeed gained. Identical reasoning led to the development of the first colour television demonstrations by Baird in 1928, and the first public colour television transmissions in America by CBS in 1951. Known as a field-sequential system, in essence the apparatus consisted of a high field-rate monochrome television system with optical red, green and blue filters presented in front of the camera lens and the receiver screen which, when synchronized, produced a colour picture. Such an electromechanical system was not only unreliable and cumbersome, but also required three times the bandwidth of a monochrome system (because three fields had to be reproduced in the period previously taken by one). In fact, even with the high field-rate adopted by CBS, the system suffered from colour flicker on saturated colours, and was soon abandoned after transmissions started.

This sequential technique had, however, a number of distinct advantages – high brightness and perfect registration between the three coloured images among them. These strengths alone have kept the idea of sequential displays alive for 50 years. The American company Tektronix has very recently developed a modern version of this system, which employs large format liquid-crystal shutter technology (LCS). The LCS is an electronic switchable colour filter that employs combinations of colour and neutral polarizers that split red, green and blue light from a specially selected monochrome CRT into orthogonally polarized components. Each component is then 'selected' by the action of a liquid crystal cell. For the engineers working in the 1950s, LCDs were, of course, not an option. Instead, they took the next most obvious logical step for producing coloured images. They argued that rather than presenting sequential fields of primary colours, present sequential dots of each primary. Such a (dot sequential) system using the secondary primaries of yellow, magenta, cyan and black forms the basis of colour printing. In a television system, individual phosphor dots of red, green and blue – provided they are displayed with sufficient spatial frequency – provide the impression of a colour image when viewed from a suitable distance.

Consider the video signal designed to excite such a dot sequential tube face. When a monochrome scene is being displayed, the television signal does not differ from its black and white counterpart. Each pixel (of red, green and blue) is equally excited, depending on the overall luminosity (or luminance) of a region of the screen. Only when a colour is reproduced does the signal start to manifest a high-frequency component related to the spatial frequency of the phosphor it's designed successively to stimulate. The exact phase of the high-frequency component depends, of course, on which phosphors are to be stimulated. The more saturated the colour (i.e. the more it departs from grey), the more high-frequency 'colourizing' signal is added. This signal is mathematically identical to a black and white television signal whereupon is superimposed a high-frequency colour-information carrier signal (now known as a colour subcarrier) – a single-frequency carrier whose instantaneous value of amplitude and phase respectively determines the saturation and hue or any particular region of the picture. This is the essence of the NTSC colour television system launched in the USA in 1953 although, for practical reasons, the engineers eventually resorted to an electronic dot sequential signal rather than achieving this in the action of the tube. This technique is considered next.

NTSC and PAL colour systems

If you've ever had to match the colour of a cotton thread or wool, you'll know you have to wind a length of it around a piece of card before you are in a position to judge the colour. That's because the eye is relatively insensitive to coloured detail. This is obviously a phenomenon of great relevance to any application of colour picture reproduction and coding; that colour information may be relatively coarse in comparison with luminance information. Artists have known this for thousands of years. From cave paintings to modern animation studios, it's possible to see examples of skilled, detailed monochrome drawings being coloured in later by a less skilled hand.

The first step in the electronic coding of an NTSC colour picture is colour–space conversion into a form where brightness information (luminance) is separate from colour information (chrominance), so that the latter can be used to control the high-frequency colour subcarrier. This axis transformation is usually referred to as RGB to YUV conversion, and it is achieved by mathematical manipulation of the form:

$$Y = 0.3R + 0.59G + 0.11B$$

$$U = m(B - Y)$$

$$V = n(R - Y)$$

The Y (traditional symbol for luminance) signal is generated in this way so that it as nearly as possible matches the monochrome signal from a black and white camera scanning the same scene (the colour green is a more luminous colour than either red or blue, and red is more luminous than blue). Of the other two signals, U is generated by subtracting Y from B; for a black and white signal this evidently remains zero for any shade of grey. The same is true of $R - Y$. These signals therefore denote the amount a colour signal differs from its black and white counterpart, and they are dubbed colour difference signals. (Each colour difference signal is scaled by a constant.) These signals may be a much lower bandwidth than the luminance signal because they carry colour information only, to which the eye is relatively insensitive. Once derived, they are low-pass filtered to a bandwidth of 0.5 MHz. These two signals are used to control the amplitude and phase of a high-frequency subcarrier superimposed onto the luminance signal. This chrominance modulation process is implemented with two balanced modulators in an amplitude-modulation suppressed-carrier configuration – a process that can be thought of as multiplication. A clever technique is employed so that U modulates one carrier signal and V modulates another carrier of identical frequency, but phase-shifted with respect to the other by ninety degrees. These two carriers are then combined and result in a subcarrier signal, which varies its phase and amplitude dependant upon the instantaneous value of U and V. Note the similarity between this and the form of colour information noted in connection with the dot sequential system: amplitude of high-frequency carrier dependant upon the depth – or saturation – of the colour, and phase dependant upon the hue of the colour. (The difference is that in NTSC, the colour subcarrier signal is coded and decoded using electronic multiplexing and de-multiplexing of YUV signals rather than the spatial multiplexing of RGB components attempted in dot sequential systems.) Figure 2.13 illustrates the chrominance coding process.

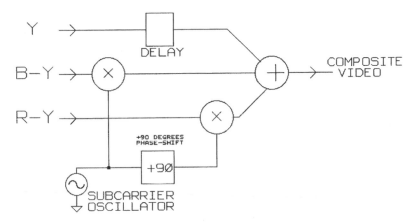

Figure 2.13 *NTSC coder block schematic*

Whilst this simple coding technique works well, it suffers from a number of important drawbacks. One serious implication is that if the high-frequency colour subcarrier is attenuated (for instance due to the low pass action of a long coaxial cable), there is a resulting loss of colour saturation. More serious still, if the phase of the signal suffers from progressive phase disturbance, the colour in the reproduced colour is likely to change. This remains a problem with NTSC, where no means are taken to ameliorate the effects of such a disturbance. The PAL system takes steps to prevent phase distortion having such a disastrous effect by switching the phase of the V subcarrier on alternate lines. This really involves very little extra circuitry within the coder, but has design ramifications that mean the design of PAL decoding is a very complicated subject indeed. The idea behind this modification to the NTSC system (for that is all PAL is) is that, should the picture – for argument's sake – take on a red tinge on one line, it is cancelled out on the next when it takes on a complementary blue tinge. The viewer, seeing this from a distance, just continues to see an undisturbed colour picture. In fact, things aren't quite that simple because Dr Walter Bruch (who invented the PAL system) was obliged to use an expensive line-delay element and a following blend circuit to effect an electrical cancellation – rather than a pure optical one. Still, the concept was important enough to be worth naming the entire system after this one notion – phase alternation line (PAL). Another disadvantage of the coding process illustrated in Figure 2.13 is due to the contamination of luminance information with chrominance, and vice versa. Although this can be limited to some degree by complementary band-pass and band-stop filtering, a complete separation is not possible, and this results in the swathes of moving coloured bands (cross-colour) that appear across high-frequency picture detail on television – herring-bone jackets proving especially potent in eliciting this system pathology.

In the colour receiver, synchronous demodulation is used to decode the colour subcarrier. One local oscillator is used, and the output is phase shifted to produce the two orthogonal carrier signals for the synchronous demodulators (multipliers). Figure 2.14 illustrates the block schematic of an NTSC colour decoder. A PAL decoder is much more complicated.

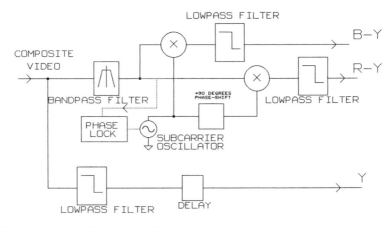

Figure 2.14 *NTSC decoder block schematic*

Mathematically, we can consider the coding and decoding process thus:

$$\text{NTSC colour signal} = Y + 0.49(B - Y)\sin \omega t + 0.88(R - Y)\cos \omega t$$

$$\text{PAL colour signal} = Y + 0.49(B - Y)\sin \omega t \pm 0.88(R - Y)\cos \omega t$$

Note that, following the demodulators, the U and V signals are low-pass filtered to remove the twice frequency component, and the Y signal is delayed to match the processing delay of the demodulation process before being combined with the U and V signals in a reverse colour space conversion. In demodulating the colour subcarrier, the regenerated carriers must not only remain spot-on frequency, but also maintain a precise phase relationship with the incoming signal. For these reasons the local oscillator must be phase-locked, and for this to happen the oscillator must obviously be fed a reference signal on a regular and frequent basis. This requirement is fulfilled by the colour burst waveform, which is shown in the composite colour television signal displayed in Figure 2.15. The

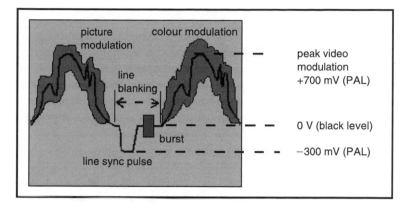

Figure 2.15 *Composite colour TV signal (at line rate)*

reference colour burst is included on every active television line at a point in the original black and white signal given over to line retrace. Notice also the high-frequency colour information superimposed on the 'black and white' luminance information. Once the demodulated signals have been through a reverse colour space conversion and become RGB signals once more, they are applied to the guns of the colour tube.

Further details of the NTSC and PAL systems are shown in Table 2.2.

Table 2.2

	NTSC	*PAL*
Field frequency	59.94 Hz	50 Hz
Total lines	525	625
Active lines	480	575
Horizontal resolution	440	572
Line frequency	15.75 kHz	15.625 kHz

(Note: Horizontal resolutions calculated for NTSC bandwidth of 4.2 MHz and 52 µs line period; PAL, 5.5 MHz bandwidth and 52 µs period.)

SECAM colour system

In 1961, Henri de France put forward the SECAM system (Sequentiel Couleur a Memoire) in which the two chrominance components (U and V) are transmitted in sequence, line after line, using frequency modulation. In the receiver, the information carried in each line is memorized until the next line has arrived, and then the two are processed together to give the complete colour information for each line.

Shadowmask tube

As you watch television, three colours are being scanned simultaneously by three parallel electron beams, emitted by three cathodes at the base of the tube and all scanned by a common magnetic deflection system. But how to ensure that each electron gun only excites its appropriate phosphor? The answer is the shadowmask – a perforated, sheet-steel barrier that masks the phosphors from the action of an inappropriate electron gun. The arrangement is illustrated in Figure 2.16. For a colour tube to produce an acceptable picture at reasonable viewing distance, there are about half a million phosphor red, green and blue triads on the inner surface of the screen. The electron guns are set at a small angle to each other and aimed so that they converge at the shadowmask. The beams then pass through one hole and diverge a little between the shadowmask and the screen so that each strikes only its corresponding phosphor.

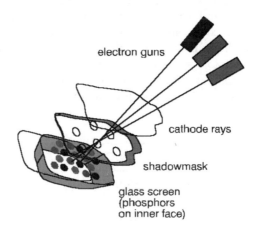

electron guns

cathode rays

shadowmask

glass screen
(phosphors
on inner face)

Figure 2.16   *Action of the shadowmask in colour tube*

Waste of power is one of the very real drawbacks of the shadowmask colour tube. Only about a quarter of the energy in each electron beam reaches the phosphors. Up to 75 per cent of the electrons do nothing but heat up the steel! For a given beam current, a colour tube is very much fainter than its monochrome counterpart. In a domestic role such inefficiency is of little importance, but this is not so when CRT's are called upon to do duty in head-mounted displays or to act as colour displays in aeroplane cockpits, or indeed to perform anywhere power is scarce or heat is damaging. These considerations have caused manufacturers to re-visit dot sequential colour systems. These implementations incorporate CRT's known as beam index tubes. Essentially, the tube face in a beam index tube is arranged in vertical phosphor strips. As noted above, dot sequential systems rely on knowing the exact position of the scanning electron beam; only in this way can the appropriate phosphor be excited at the appropriate time. This is achieved in the beam index tube by depositing a fourth phosphor at intervals across the screen face that elicits an ultraviolet light when energized by the electron beam. As the beam scans across the tube face, a series of ultraviolet pulses is emitted from the phosphor strips. These pulses are detected by way of an ultraviolet photo detector mounted on the flare of the tube. These pulses are intercepted and turned into electrical pulses, and these are used within a feedback system to control the multiplexer, which sequentially selects the three colour signals to the single electron gun.

Vestigial sideband modulation

In order to transmit television signals by radio techniques, the final television signal is made to amplitude-modulate an RF carrier wave. Normally modulation is arranged so that the tips of the sync pulse represent the minimum modulation (maximum carrier), with increasing video signal amplitude causing a greater and greater suppression of carrier (as shown in Figure 2.17). This modulation scheme is termed negative video modulation.

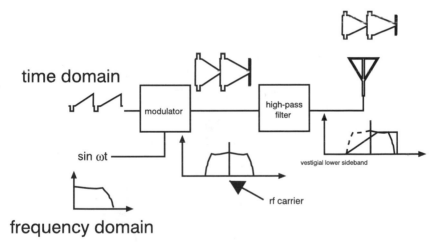

Figure 2.17 *Vestigial sideband modulation*

The bandwidth of an analogue television signal is in the region of 4–6 MHz, depending on the system in use. If this were used to amplitude-modulate a RF carrier, as in sound broadcasting, the radio bandwidth would be twice this figure. Adding space for a sound carrier and allowing sufficient gaps in the spectrum so that adjacent channels would not interfere with each other would result in an unacceptably broad overall RF bandwidth and a very inefficient use of the radio spectrum. To this end, it was found that AM modulation and reception works almost as well if one sideband is filtered out at the transmitter, as shown in Figure 2.17; provided that the tuned circuits in the receiver are appropriately tuned to accept only one sideband too. All that is transmitted is the carrier one sideband, the other being more or less completely filtered off. This filtered sideband is thus termed a vestigial sideband. Looking at the modulated signal, even with one sideband missing, amplitude modulation still exists and, when the RF signal is rectified, the modulation frequency is still extracted.

Audio for television

Analogue audio for television is always carried by a RF carrier adjacent to the main vision carrier, in the 'space' between the upper sidebands of the vision carrier and the vestigial lower sideband of the next channel in the waveband, as illustrated in Figure 2.18. The precise spacing of the audio carrier from the vision carrier varies from system to system around the world, as does the modulation technique. Some systems use an amplitude modulated carrier; others an FM system. Some stereo systems use two carriers, the main carrier to carry a mono (L + R) signal and the secondary

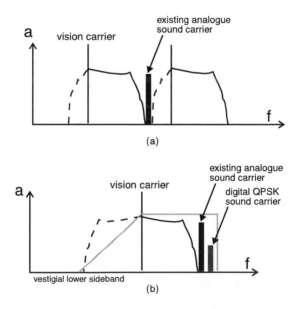

Figure 2.18 (a) *Sound carrier – relation to vision carriers;* (b) *Position of digital sound carrier*

carrier to carry $(L - R)$ information; in this way compatibility with non-stereo receivers is maintained. Current stereo audio systems in analogue television also use pure digital techniques.

NICAM 728 digital stereo sound

Real audio signals do not change instantaneously from very large to very small values, and even if they did we would not hear it due to the action of temporal masking described earlier. So a form of digital temporal-based signal compression may be applied. This is the principle behind the stereo television technique of NICAM, which stands for Near Instantaneous Companded Audio Multiplex. (NICAM is explained in Chapter 6.) The final digital signal is coded onto a low-level carrier just above the analogue sound carrier (see Figure 2.18). The carrier is modulated using differential quadrature phase-shift keying (or DQPSK, see Chapter 10).

Recording television signals

Even before the days of colour television, video tape recorders (VTR) had always been faced with the need to record a wide bandwidth signal because a television signal extends from DC to perhaps 4 or 5 MHz (depending on the system). The DC component in an analogue television signal exists to represent overall scene brightness.

There exist two fundamental limitations to the reproducible bandwidth from an analogue tape recorder of the type considered so far. The first is due to the method of induction of an output signal; which is – in turn – due to the rate of change of flux in the tape head. Clearly a zero frequency signal can never be recorded and reproduced because, by definition, there would exist no change in flux and therefore no output signal. In fact, the frequency response of an unequalized tape recorder varies linearly with frequency; the higher the frequency, the faster the rate of change of flux and the higher the induced electrical output. This effect is illustrated in Figure 2.19. In audio tape recorders the intrinsic limitation of an inability

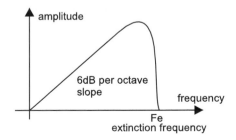

Figure 2.19 *Amplitude response of magnetic tape recording system*

to record zero frequency is not important because, usually, 20 Hz is regarded as the lowest frequency required to be reproduced in an audio system. Similarly, the changing frequency response is 'engineered around' by the application of complementary equalization. But in video tape recorders, where the DC component must be preserved, this is achieved by the use frequency modulation; a modulation scheme in which a continuous modulating frequency is present at the tape heads even if there is little or no signal information, or where the signal information changes very slowly.

The second limitation is a function of recorded wavelength and head gap. Essentially, once the recorded wavelength approaches the dimension of the head gap, the response of the record–replay system falls sharply as is illustrated in Figure 2.19. (The response reaches total extinction at the frequency at which the recorded wavelength is equal to that of the head gap.) Of course, recorded wavelength is itself a function of linear tape speed – the faster the tape travels, the longer the recorded wavelength – so theoretically the bandwidth of a tape recorder can be extended indefinitely by increasing the tape speed. It's pretty clear, however, that there are some overwhelming practical and commercial obstacles to such an approach.

The alternative approach developed first by Ampex in the VR-1000 video tape recorder was to spin a number of heads in a transverse fashion across the width of the tape, thereby increasing the head-to-tape writing speed without increasing the linear tape speed. This video technology was named Quadruplex, after the four heads that rotated on a drum across a 2" wide tape. Each video field was written in one rotation of the drum, so each video field was split into four sections. This led to one of the problems which beset 'Quad', as this tape format is often called, where the picture appears to be broken into four discrete bands due to differing responses from each of the heads. During the 1950s many companies worked on variations of the Ampex scheme that utilized the now virtually universal helical recording format; a scheme (illustrated in Figure 2.20) in which the tape is wrapped around a rotating drum that contains just two-heads. One head writes (or reads) a complete field of video in a slanting path across the width of the head. By this means head switching can be made to happen invisibly, just before a vertical-blanking interval. Virtually all video tape formats (and contemporary digital audio tape recording formats) employ a variation of this technique.

Colour under

This then was the basis of a monochrome video tape recorder; a spinning head arrangement writing an FM modulated TV signal across the tape in

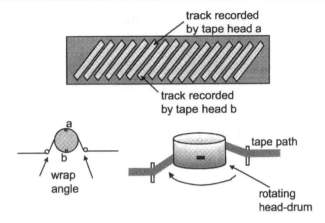

Figure 2.20 *Helical scan arrangement*

'sweeps' at field rate. Although such a system can be used for a composite (colour) television signal, and was, in practice it proved hard to make this technology work without a considerable technological overhead – an acceptable situation in professional broadcast but unsuitable for low-cost recorders for industrial and home use. The problem lay in the colour information, which is phase modulated on a precise subcarrier frequency. Any modulation of the velocity of the head as it swept the tape resulted in a frequency change in the chroma signal off tape. This produced very great colour signal instability without the use of expensive time-base correctors (TBCs) in the playback electronics.

The solution came with the development of the colour-under technique. In this system, colour and luminance information is separated and the former is superheterodyned down to a frequency below the FM modulated luminance signal (the so-called *colour-under* signal). It is this 'pseudo-composite' signal (illustrated in Figure 2.21) that is actually written onto tape. Very cleverly, the new colour IF carrier recorded onto tape is created

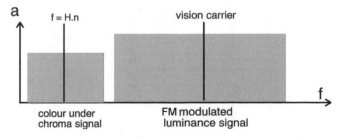

Figure 2.21 *Spectrum of vision carrier modulation and colour-under signal*

by an oscillator that runs at a multiple of the line frequency. In replay –
provided this local oscillator is kept in phase lock with the horizontal syncs
derived from the luminance playback electronics – a stable subcarrier
signal is recovered when it is mixed with the off-tape colour-under signal.

Audio tracks

Although there have been other systems for analogue audio involving FM
carriers recorded onto tape with the spinning video tape head, analogue
audio is classically carried by linear tracks that run along the edge of the
video tape in the region where the spinning head is prevented from
writing for fear of snagging the tape edge. Often there are a number of
these tracks, for stereo etc. One of these linear tracks is often reserved for
the quasi-audio signal known as timecode, which is described below.

Timecode

Longitudinal timecode

Timecode operates by 'tagging' each video frame with a unique identify-
ing number called a timecode address. The address contains information
concerning hours, minutes, seconds and frames. This information is
formed into a serial digital code, which is recorded as a data signal onto
one of the audio tracks of a video tape recorder. (Some video tape
recorders have a dedicated track for this purpose.) Each frame's worth of
data is known as a word of timecode, and this digital word is formed of 80
bits spaced evenly throughout the frame. Taking EBU timecode[2] as an
example, the final data rate therefore turns out to be 80 bits × 25 frames
per second = 2000 bits per second, which is equivalent to a fundamental
frequency of 1 kHz; easily low enough, therefore, to be treated as a
straightforward audio signal. The timecode word data-format is illustrated
(along with its temporal relationship to a video field) in Figure 2.22. The
precise form of the electrical code for timecode is known as Manchester
bi-phase modulation. When used in a video environment, timecode must
be accurately phased to the video signal. As defined in the specification,
the leading edge of bit '0' must begin at the start of line 5 of field 1 (±1
line). Time address data is encoded within the 80 bits as eight 4-bit BCD
(binary coded decimal) words (i.e. one 4-bit number for tens and one for
units). Like the clock itself, time address data is only permitted to go from:
00 hours, 00 minutes, 00 seconds, 00 frames to 23 hours, 59 minutes, 59
seconds, 24 frames.

   However, a 4-bit BCD number can represent any number from 0 to 9, so
in principle timecode could be used to represent 99 hours, 99 minutes and
so on. But as there are no hours above 23, no minutes or seconds above

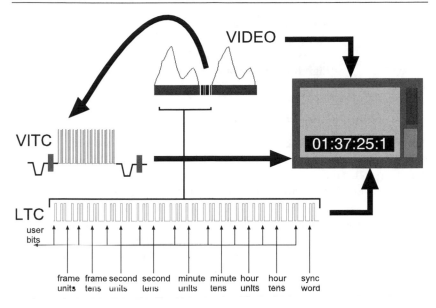

Figure 2.22 *Format of LTC and VITC timecode*

59 and no frames above 24 (in PAL), timecode possesses potential redundancy. In fact, some of these extra codes are exploited in other ways. The basic time address data and these extra bits are assigned their position in the full 80-bit timecode word like this:

0–3	Frame units
4–7	First binary group
8–9	Frame tens
10	Drop frame flag
11	Colour frame flag
12–15	Second binary group
16–19	Seconds units
20–23	Third binary group
24–26	Seconds tens
27	Unassigned
28–31	Fourth binary group
32–35	Minutes units
36–39	Fifth binary group
40–42	Minutes tens
43	Unassigned
44–47	Sixth binary group
48–51	Hours units

52–55	Seventh binary group
56–57	Hours tens
58–59	Unassigned
60–63	Eighth binary group
64–79	Synchronizing sequence

Vertical interval timecode (VITC)

Longitudinal timecode (LTC) is a quasi-audio signal recorded on an audio track (or hidden audio track dedicated to timecode). VITC, on the other hand, encodes the same information within the vertical interval portion of the video signal in a manner similar to a Teletext signal (see below). Each has advantages and disadvantages; LTC is unable to be read while the player/recorder is in pause, while VITC cannot be read whilst the machine is in fast forward or rewind modes. It is advantageous that a video tape should have both forms of timecode recorded. VITC is illustrated in Figure 2.22 too. Note how timecode is displayed 'burned-in' on the monitor.

PAL and NTSC

Naturally timecode varies according to the television system used, and for NTSC there are two versions of timecode in use to accommodate the slight difference between the nominal frame rate of 30 frames per second and the actual frame rate of NTSC of 29.97 frames per second. While every frame is numbered and no frames are ever actually dropped, the two versions are referred to as 'drop'- and 'non-drop'-frame timecode. Non-drop-frame timecode will have every number for every second present, but will drift out of relationship with clock time by 3.6 seconds every hour. Drop-frame timecode drops numbers from the numbering system in a predetermined sequence, so that the timecode-time and clock-time remain in synchronization. Drop-frame is important in broadcast work, where actual programme time is important.

User bits

Within the timecode word there is provision for the hours, minutes, seconds, frames and field ID that we normally see, and for 'user bits', which can be set by the user for additional identification. Use of user bits varies, with some organizations using them to identify shoot dates or locations and others ignoring them completely.

Teletext™

Teletext was first introduced in the UK in 1976, and grew out of research to provide captions for the deaf and for foreign language subtitles. The former still remains one of the most important and worthwhile functions of Teletext. The UK Teletext (BBC Ceefax™ or ITV Oracle™) system is described below. Details differ, but all services of this type use essentially the same techniques. For instance, the French system (Antiope™) is capable of richer accented text, is not compatible with Teletext, but shares practically all the same engineering principles although the coding techniques are more complicated.

Teletext uses previously unused lines in the field-blanking interval to broadcast a digital signal, which represents pages of text and graphical symbols. Each transmitted character is coded as a 7-bit code together with a single parity bit to form an 8-bit byte. A page of Teletext contains 24 rows of 40 characters, and a row of characters is transmitted in each Teletext line. Together with 16 data clock run-in bits, an 8-bit frame coding byte and two 8-bit bytes carrying row-address and control information, this makes a final signal data-rate of nearly 7 Mbits/s, which is accommodated in the 5 MHz video bandwidth of a PAL channel. Note that the system does not transmit text in a graphical manner – it is in the receiver that the bitmap text is generated, although there is provision for a rather limited graphics mode that is used for headlines and banners.

The top row of each page is actually a page header, and contains page address information, time of day clock and date. This page is used to establish the start of a page to be stored in memory. When the page address matches the page number requested by the viewer, the Teletext decoder will display the page on screen. The system allows for eight 'magazines' of 100 pages each. The position of Teletext lines in the field interval is shown in Figure 2.23, and form of a typical data line is also shown. As we have seen, each television line is used to transmit one row of characters. If two lines in each field are used, it takes 12 fields (or nearly a quarter of a second) to transmit each page and 24 seconds to transmit a whole magazine of 100 pages. If each page was transmitted in sequence this would result in rather a long wait – and in fact this is the case with many Teletext pages – but this is partially resolved by transmitting the most commonly used pages (the index, for example) more often.

Analogue high definition television (HDTV)

Before the really incredible advances made in source coding (data compression) techniques for television pictures, there was great and widespread pessimism concerning the ability to transmit digital signals direct to the home (by terrestrial, satellite or cable techniques). This was all the more disappointing because during the 1980s interest started growing in 'better' television with a wider aspect ratio and sharper pictures; something at least approaching watching a cinema film.

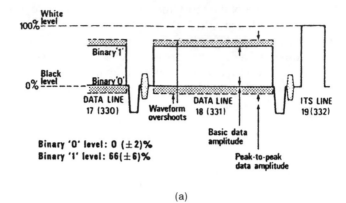

(a)

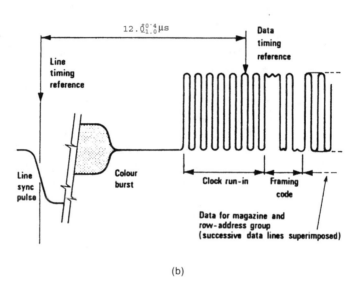

(b)

Figure 2.23　*Position of data line and coding for Teletext information*

The display of PAL and NTSC pictures is practically (although not theoretically) bandwidth-limited in the luminance domain to 3.0–3.5 MHz, depending on the system and the complexity of the decoder. This does not give exceptional quality on a 4 : 3 receiver, and in 16 : 9 each pixel is effectively stretched by 30 per cent – leading to even worse definition. This, coupled with cross-colour and cross-luminance effects, all of which are inevitably exaggerated by a large screen, led to the development of several attempts to 'improve' Gerber system 625/50 (PAL) pictures in order to give a near-film experience but using analogue or hybrid analogue–digital techniques.

One such attempt that had some commercial success was MAC, or multiplex analogue components; another, PALplus.

MAC

The starting point of a MAC television signal was the separation of luminance and colour information using a time-division multiplex rather than the frequency-division multiplex used in PAL. By this means a very much better separation can be ensured. In a MAC television signal, the rather higher bandwidth (anamorphic 16:9) luminance signal is squashed into 40 μs of the 52 μs active line; followed by a squashed chroma component. In the following line the second chroma component is transmitted (a bit like SECAM). The receiver is responsible for 'un-squashing' the line back to 52 μs, and for combining the luminance and the chrominance. There exist several versions of MAC, the most common being D-MAC; most of the variants concern the coding of the sound signal. This is usually accomplished with digital techniques.

PALplus

A PALplus picture appears as a 16:9 letterboxed image with 430 active lines on conventional 4:3 TVs, but a PALplus equipped TV will display a 16:9 picture with the full 574 active lines. A PALplus receiver 'knows' it is receiving a PALplus signal due to a data burst in line 23, known as the wide screen signalling (WSS) line.

Inside the PALplus TV the 430 line image is 'stretched' to fill the full 574 lines, the extra information needed to restore full vertical resolution being carried by a subcarrier-like 'helper' signal coded into the letterbox black bars. Horizontal resolution and lack of cross-colour artefacts is assured by the use of the Clean PAL encoding and decoding process, which is essentially an intelligent coding technique that filters luminance detail which is particularly responsible for cross-colour effects.

Some of the bits of the WSS line data tell a PALplus decoder whether the signal originated from a 50-Hz interlaced camera source or from 25 fps film via a telecine machine. PALplus receivers have the option of de-interlacing a 'film mode' signal and displaying it on a 50-Hz progressive-scan display (or using ABAB field repeat on a 100-Hz interlaced display) without risk of motion artefacts. Colour decoding parameters too are adjusted according to the mode in use.

1125/60 and 1250/50 HDTV systems

In Japan, where the NTSC system is in use, pressure to improve television pictures resulted in a more radical option. In part this was due to the bandwidth limitations of the NTSC system. These are even more severe than PAL (525 line NTSC was, after all, invented in 1939) and attempts to 're-engineer' NTSC proved very difficult. Instead, research in NHK (the Japanese state broadcast organization) resulted in a recommendation for a new high-definition television system using a 1125, 60-Hz field interlace scanning system with a 16:9 aspect ratio screen and a luminance bandwidth of 30 MHz! This is evidently too large for sensible spectrum use, and the Japanese evolved the MUSE system for transmitting HDTV. MUSE is a hybrid analogue–digital system which sub-samples the HDTV picture in four passes. The result is excellent on static pictures, but results in loss of definition on moving objects.

1250/50 European HDTV

In high dudgeon, some European television manufacturers decided to develop a European equivalent of the Japanese 1125/60 production system. This utilized 1250 lines interlaced in two 50-Hz fields. It has not been widely adopted.

625-line television wide screen signalling

For a smooth introduction of new television services with a 16:9 display aspect ratio in PAL and SECAM standards, it is necessary to signal the aspect ratio to the television receiver. The wide screen signalling (WSS) system (European Standard: EN 300 294 V1.3.2, 1998-04) standardizes this signalling information; the idea being that the receiver should be capable of reacting automatically to this information by displaying the video information in a specified aspect ratio. The signalling described here is applicable for 625-line PAL and SECAM television systems.

The WSS signal

The signalling bits are transmitted as a data burst in the first part of line 23, as illustrated in Figure 2.24. Each frame line 23 carries the WSS.

Data structure

The WSS signal has a three-part structure: Preamble, Data Bits, and Parity Bit. The preamble contains a run-in and a start code. This is followed by 14 data bits (bits 0–13). The data bits are grouped in 4 data groups, as

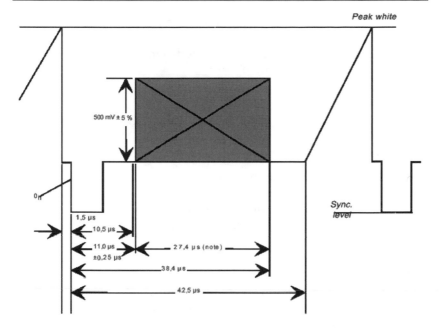

Figure 2.24 *Position of status bit signalling in line 23*

explained below. For error detection, the WSS finishes with an odd parity bit. The four groups of data bits are now considered.

Group 1: *Aspect ratio label, letterbox and position code*

Bits 0–3	Aspect ratio label	Format	Position	No. of active lines
0001	4:3	full format	not applicable	576
1000	14:9	letterbox	centre	504
0100	14:9	letterbox	top	504
1101	16:9	letterbox	centre	430
0010	16:9	letterbox	top	430
1011	>16:9	letterbox	centre	not defined
0111	14:9	full format	centre	576
1110	16:9	anamorphic	not applicable	576

Group 2: *Enhanced services*

Bit 4: Denotes the film bit (0 camera mode: 1 film mode). This can be used to optimize decoding functions; because film has 2 identical fields in a frame.

Bit 5: Denotes the colour coding bit (0 = standard coding: 1 = Motion Adaptive Colour Plus).
Bit 6: Helper bit (0 No helper: 1 Modulated helper).
Bit 7: Reserved (always set to '0').

Group 3:   *Subtitles*

Bit 8: Flags that subtitles are contained within Teletext (0 no subtitles within Teletext: 1 subtitles within Teletext)
Bits 9 and 10: Subtitling mode

Bits 9–10	Subtitles and position
0 0	no open subtitles
1 0	subtitles in active image area
0 1	subtitles out of active image area
1 1	reserved

Group 4:   *Others*

Bit 11: Denotes the surround sound (0 no surround sound information: 1 surround sound mode)
Bit 12: Denotes the copyright and generation (0 no copyright asserted or status unknown: 1 copyright asserted)
Bit 13: Generation bit (0 copying not restricted: 1 copying restricted)

Display formats

The WSS standard (EN 300 294 V1.3.2, 1998–04) also specifies the minimum requirements for the method of display for material formatted in each of the different aspect ratios with the proviso that the viewer should always be free to override the automatically selected display condition. It should be evident that the speed of the automatic change of aspect ratio is limited mainly by the response time of the TV deflection circuit!

Telecine and 'pulldown'

As we have already seen, an NTSC video image consists of 525 horizontal lines of information; a PAL signal consists of 625 lines of information. In either case, the electron gun scans half the lines twice over, in a process known as interlace. Each full scan of even-numbered lines, or odd-numbered lines, constitutes a 'field' of video. In NTSC, each field scan takes approximately 1/60th of a second; therefore a whole frame is scanned each 1/30th of a second. Actually, due to changes made in the 1950s to accommodate the colour information, the NTSC frame-rate is not exactly 30 frames-per-second (fps) but is 29.97 fps. In PAL, each field scan takes exactly 1/50th of a second and the frame-rate is exactly 1/25th of a second. Unfortunately neither is compatible with the cinema film frame-rate.

Film is generally shot and projected at 24 fps, so when film frames are converted to video, the rate must be modified to play at 25 fps (for PAL) or 29.97 fps (for NTSC). How this is achieved is very different, depending on whether the target TV format is PAL or NTSC. In PAL, the simple technique is employed of running the cinema film slightly (25/24 or 4%) fast. This clearly has implications for running time and audio pitch, but the visual and audible distortion is acceptable. Not so NTSC, where running the film 30/40 or 25% fast would be laughable, so a different technique is required. This technique is known as 'pulldown'.

Film is transferred to video in a machine known as a telecine machine. The technology differs but, in each case, film is drawn in a projector-like, mechanical, intermittent advance mechanism in front of a lens and electronic-scanning arrangement. In PAL, the only complication is that the mechanical film-advance must be synchronized with the video signal scanning. For conversion to NTSC, the telecine must perform the additional 'pulldown' task associated with converting 24 frames of film to 30 frames of video. In the pulldown process, twelve extra fields (equivalent to 6 frames) are added to each 24 frames of film, so that the same images that made up 24 frames of film then comprise 30 frames of video. (Remember that real NTSC plays at a speed of 29.97 fps, so the film actually runs at 23.976 fps when transferred to video.)

If we imagine that groups of four frames of film numbered A_n, B_n, C_n, D_n (where n increments with each group), pulldown performs this generation of extra frames like this: the first frame of video contains two fields, scanned from one frame of film (A). The second frame of video contains two fields scanned from the 2nd frame of film (B). The third frame of video contains one field scanned from the 2nd (B) and one field scanned from the 3rd frame of film (C). The fourth frame of video contains one field scanned from the 3rd (C) and one filed from the 4th frames of film (D). The fifth frame of video contains two fields scanned from the 4th (D) frame of film. We could represent this process like this:

A1/A1, B1/B1, B1/C1, C1/D1, D1/D1, A2/A2, B2/B2, B2/C2, C2/D2, D2/D2, A3/A3 ...

Note the pattern 3,2,3,2. For this reason, this particular technique is called '2–3 pulldown' (there's an alternative, with a reversed pattern, called '3–2 pulldown'). Either way, four frames of film become five frames of video in which is what's required because

$$30/24 \text{ (or } 29.97/23.976) = 5/4$$

Despite the motion artefacts which this intermittent and irregular mapping creates, a pulldown NTSC dub of a film has the advantage that it runs at nearly exactly the right speed (0.1% slow), which makes it more suitable for post-production work; for example, as a working tape for film music editor and composer. All the above is nicely distilled and captured in Figure 2.25.

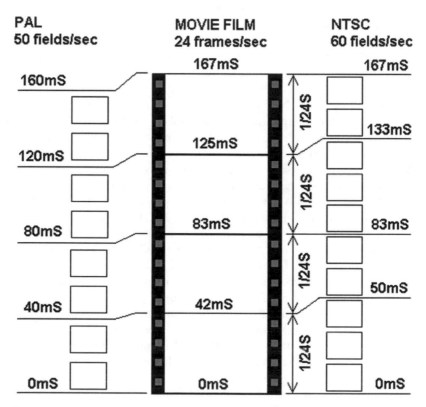

Figure 2.25   *A clear illustration of the transfer of film to TV in the NTSC and PAL world*

Reference

Moore, B.C.J. (1989) *An Introduction to the Psychology of Hearing* (3rd edition). Academic Press.

Notes

1 The phase of a sine wave oscillation relates to its position with reference to some point in time. Because we can think of waves as cycles, we can

express the various points on the wave in terms of an angle relative to the beginning of the sine wave (at 0°). The positive zero crossing is therefore at 0°, the first peak at 90° etc.

2 European Broadcasting Union (EBU) timecode is based on a field frequency of 25 frames per second.

3
Digital video and audio coding

Digital fundamentals

In order to see the forces that led to the rapid adoption of digital video processing and interfacing throughout the television industry in the 1990s, it is necessary to look at some of the technical innovations in television during the late 1970s and early 1980s.

The NTSC and PAL television systems described previously were primarily developed as transmission standards, not as television production standards. As we have seen, because of the nature of the NTSC and PAL signal, high-frequency luminance detail can easily translate to erroneous colour information. In fact, this cross-colour effect is an almost constant feature of the broadcast standard television pictures, and results in a general 'busyness' to the picture at all times. That said, these composite TV standards (so named because the colour and luminance information travel in a composite form) became the primary production standard, mainly due to the inordinate cost of 'three-level' signal processing equipment (i.e. routing switchers, mixers etc.), which operated on the red, green and blue or luminance and colour-difference signals separately. A further consideration, beyond cost, was that it remained difficult to keep the gain, DC offsets and frequency response (and therefore delay) of such systems constant, or at least consistent, over relatively long periods of time. Systems that did treat the signal components separately suffered particularly from colour shifts throughout the duration of a programme. Nevertheless, as analogue technology improved with the use of integrated circuits as opposed to discrete semiconductor circuits, manufacturers started to produce three-channel, component television equipment which processed the luminance, $R - Y$ and $B - Y$ signals separately. Pressure for this extra quality came particularly from graphics, where people found that working with the composite standards resulted in poor quality images that were tiring to work on, and where they wished to use both fine detail textures, which created cross-colour, and heavily saturated colours, which do not produce well on a composite system (especially NTSC).

So-called analogue component television equipment had a relatively short stay in the world of high-end production, largely because the problems of inter component levels, drift and frequency response were never ultimately solved. A digital system, of course, has no such problems. Noise, amplitude response with respect to frequency and time are immutable parameters 'designed into' the equipment, not parameters that shift as currents change by fractions of milliamps in a base emitter junction somewhere! From the start, digital television offered the only real alternative to analogue composite processing and, as production houses were becoming dissatisfied with the production value obtainable with composite equipment, the death-knell was dealt to analogue processing in television production.

However, many believed that analogue transmission methods would remain the mainstay of television dissemination because the bandwidths of digital television signals were so high. This situation has changed due to the enormous advances made in VLSI integrated circuits for video compression and decompression. These techniques will be explained in Chapter 5.

Sampling theory and conversion

There exist three fundamental differences between a continuous-time, analogue representation of a signal and a digital, pulse code modulation (PCM) description. First, a digital signal is a time-discrete, sampled representation, and secondly, it is quantized. Lastly, as we have already noted, it is a symbolic representation of this discontinuous time, quantized signal. Actually it's quite possible to have a sampled analogue signal (many exist; for instance, film is a temporally sampled system). It is also obviously quite possible to have a time-continuous, quantized system in which an electrical current or voltage could change state any time it wished but only between certain (allowed) states – the output of a multivibrator is one such circuit. The circuit that performs the function of converting a continuous-time signal with an infinite number of possible states (an analogue signal) into a binary (two state) symbolic, quantized and sampled (PCM) signal is known as an analogue-to-digital converter (ADC); the reverse process is performed by a digital-to-analogue converter (DAC).

Theory

The process of analogue-to-digital conversion and digital-to-analogue conversion is illustrated in Figure 3.1. As you can see, an early stage of

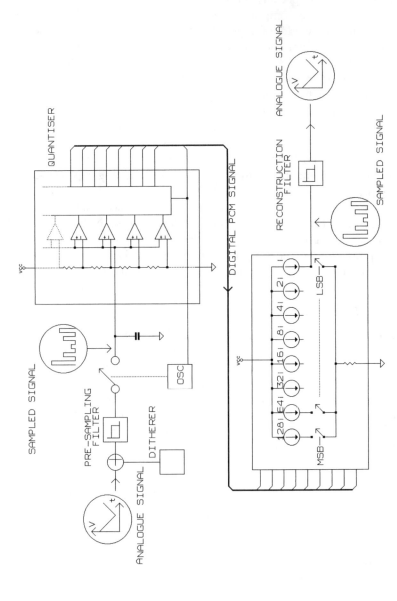

Figure 3.1 *Analogue-to-digital and digital-to-analogue conversion*

conversion involves sampling. It can be proved mathematically that all the information in a bandwidth-limited analogue signal may be sent in a series of very short, periodic 'snapshots' (samples). The rate these samples need be sent is related to the bandwidth of the analogue signal, the minimum rate required being $1/(2 \times F_t)$, where F_t represents the maximum frequency in the original signal. So, for instance, an audio signal (limited – by the filter preceding the sampler – to 15 kHz) will require pulses to be sent every $1/(2 \times 15\,000)$ seconds, or 33 microseconds.

The mechanism of sampling

Figure 3.2 illustrates the effect of sampling. Effectively, the analogue signal is multiplied (modulated) by a very short period pulse train. The spectrum of the pulse train is (if the pulses were of infinitely short period) infinite, and the resulting sampled spectrum (also shown in Figure 3.2) contains the original spectrum as well as images of the spectrum as sidebands around each of the sampling pulse harmonic frequencies. It's very important to realize the reality of the lower diagram in Figure 3.2: The signal carried in a digital system really has this spectrum. We will see when we come to discrete-time versions of Fourier analysis that all digital signals actually have this form. This, if you are of an intuitive frame of mind, is rather difficult to accept. In fact this effect is termed, even by mathematicians, the *ambiguity* of digital signals.

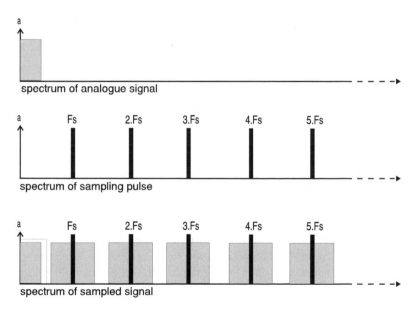

Figure 3.2   *Spectrum of a sampled signal*

Aliasing

If analogue signals are sampled at an inadequate rate it results in an effect known as aliasing, where the high frequencies get 'folded back' in the frequency domain and come out as low frequencies. Figure 3.3 illustrates the effect termed aliasing. Hence the term anti-aliasing filter for the first circuit block in Figure 3.1; to remove all frequencies above F_t.

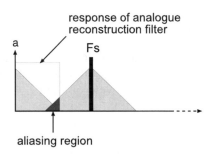

Figure 3.3   *Phenomenon of aliasing*

Quantization

After sampling, the analogue snapshots pass to the quantizer, which performs the function of dividing the input analogue signal range into a number of pre-specified quantization levels. It's very much as if the circuit measures the signal with a tape measure, with each division of the tape measure being a quantization level. The important thing to realize is that the result is always an approximation. The finer the metric on the tape measure, the better the approximations become. However, the process is never completely error-free because the smallest increment that can be resolved is limited by the accuracy and fineness of the measure. The errors may be very small indeed for a large signal, but for very small signals these errors can become discernible. This quantization error is inherent in the digital process. Some people incorrectly refer to this quantization error as quantization noise. Following the quantizer, the signal is – for the first time – a truly digital signal. However, it is often in a far from convenient form; for instance, in a video ADC the code may be a 255 parallel bit bus! So the last stage in the ADC is the code conversion, which formats the data into a binary numerical representation. The choice of the number of quantization levels determines the dynamic range of a digital PCM system. To a first approximation the dynamic range in dB is the number of digits; in the final binary numerical representation, times 6. So, an 8-bit signal has $(8 \times 6) = 48\,\mathrm{dB}$ dynamic range.

Digital-to-analogue conversion

The reverse process of digital-to-analogue conversion (also illustrated in Figure 3.1), involves regenerating the quantized voltage pulses demanded by the digital code, which may first have had to pass through a code conversion process. These pulses are then transformed back into continuous analogue signals in the block labelled reconstruction filter. The ideal response of a reconstruction filter is illustrated in Figure 3.4. This has a time-domain performance that is defined by $(\sin x)/x$. If very short pulses are applied to a filter of this type, the analogue signal is 'reconstructed' in the manner illustrated in Figure 3.5.

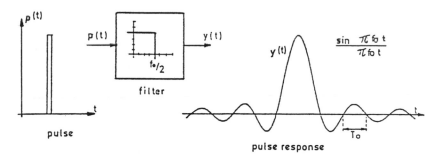

Figure 3.4 *Sin x/x impulse response of a reconstruction filter*

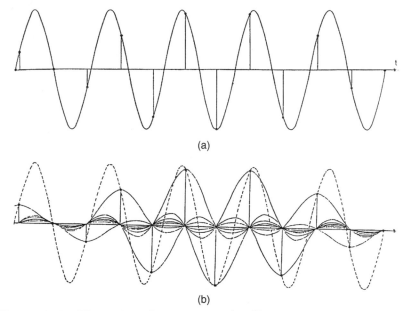

(a)

(b)

Figure 3.5 *The action of a reconstruction filter*

Jitter

There are a number of things that can adversely affect the action of sampling an analogue signal. One of these is jitter, which is a temporal uncertainty in the exact moment of sampling. On a rapidly changing signal, time uncertainty can result in amplitude quantizing errors, which in turn lead to noise. For a given dynamic range the acceptable jitter performance (expressed as a pk–pk time value) can be calculated to be:

$$\frac{\Delta T}{T_o} = \frac{1}{\pi 2^{(n-1)}}$$

where $\Delta T = $ jitter, T_o is sampling period and n is the number of bits in the system.

A few simple calculations reveal some interesting values. For instance, the equation above demonstrates that an 8-bit video signal sampled at 13.5 MHz requires that the clock jitter is not more than 92 ps; a 16-bit signal sampled at 48 kHz requires a sampling clock jitter of less than 200 ps.

Aperture effect

As we saw, the perfect sampling pulse has a vanishingly short duration. Clearly, a practical sampling pulse cannot have an instantaneous effect. The moment of sampling (t_1) is not truly instantaneous, and the converted signal doesn't express the value of the signal at t_1, but actually expresses an average value between ($t_1 - T_o/2$) and ($t_1 + T_o/2$) where (T_o) is the duration of the sampling pulse. This distortion is termed aperture effect, and it can be shown that the duration of the pulse has an effect on frequency response such that:

$$20 \log \left\{ \operatorname{sinc} \left(\frac{\pi}{2} \cdot \frac{f}{f_{o/2}} \cdot \frac{T_s}{T_o} \right) \right\} \text{dB}$$

where T_s is the maximum possible duration of the sampling pulse (aperture), $f_{o/2}$ is the Nyquist frequency limit. Note that sinc x is shorthand for $\frac{\sin x}{x}$.

Note that the aperture effect is not severe for values of $T_o < 0.2 T_s$. Even when $T_o = T_s$, the loss at the band edge (i.e. at the Nyquist frequency) is -3.9 dB. Aperture effect loss is often 'made-up' by arranging the reconstruction filter to have a compensating frequency rise.

Dither

When a quantizer converts a very large signal that crosses many quantization levels, the resulting errors from literally thousands of very slightly wrong values do indeed create a noise signal that is random in nature. Hence the misnomer quantization noise. But when a digital system records a very small signal, which only crosses a few quantization thresholds, the errors cease to be random. Instead, the errors become correlated with the signal and, because they are correlated with (or related to) the signal itself, they are far more noticeable than would be an equivalent random source of noise.

In 1984, Vanderkooy and Lipshitz proposed an ingenious and inspired answer to this problem. They demonstrated that it is possible to avoid quantization errors completely by adding a very small amount of noise to the original analogue signal prior to the analogue-to-digital converter integrated circuit. They showed that a small amount of noise is enough to break up any patterns the brain might otherwise be able to spot by shifting the signal constantly above and below the lowest quantizing thresholds. This explains the block in Figure 3.1 marked 'ditherer', which is shown as summing with the input signal prior to the sampler. In the pioneering days of digital audio and video, the design of ADCs and DACs consumed a vast amount of the available engineering effort. Today's engineer is much luckier. Many 'one-chip' solutions exist which undertake everything but a few ancillary filtering duties.

Digital video interfaces

As more and more television equipment began to process the signals internally in digital form, so the number of conversions could be kept to a minimum if manufacturers provided a digital interface standard allowing various pieces of digital video hardware to pass digital video information directly without recourse to standard analogue connections. This section is a basic outline of the 4:2:2 protocol (otherwise known as CCIR 601), which has been accepted as the industry standard for digitized component TV signals in studios for the last 10 or so years. The data signals are carried in the form of binary information coded in 8-bit or 10-bit words. These signals comprise the video signals themselves and timing reference signals (TOURS). Also included in the protocol are ancillary data and identification signals. The video signals are derived by the coding of the analogue video signal components. These components are luminance (Y) and colour difference (Cr and Cb) signals generated from primary signals (R, G, B). Cr and Cb are roughly equivalent to $R - Y$ and $B - Y$ respectively. The coding parameters are specified in CCIR Recommendation 601, and the main details are reproduced here in Table 3.1.

The sampling frequencies of 13.5 MHz (luminance) and 6.75 MHz (colour-difference) are integer multiples of 2.25 MHz, the lowest common multiple of the line-frequencies in 525/60 and 625/50 systems, resulting in a static orthogonal sampling pattern for both. The luminance

Table 3.1   *Encoding parameters values for the 4:2:2 digital video interface (10 bit values are in brackets)*

Parameters	525-line, 60 field/s systems	625-line, 50 field/s systems
1. Coded signals: Y, Cb, Cr	These signals are obtained from gamma pre-corrected RGB signals	
2. Number of samples per total line:		
luminance signal (Y)	858	864
each colour-difference signal (Cb, Cr)	429	432
3. Sampling structure	Orthogonal line, field and picture repetitive Cr and Cb samples co-sited with odd (1st, 3rd, 5th etc.) Y samples in each line	
4. Sampling frequency:		
luminance signal	13.5 MHz	13.5 MHz
each colour-difference signal	6.75 MHz	6.75 MHz

The tolerance for the sampling frequencies should coincide with the tolerance for the line frequency of the relevent colour television standard

5. Form of coding	Uniformly quantized PCM, 8 bits per sample, for the luminance signal and each colour-difference signal	
6. Number of samples per digital active line:		
luminance signal	720	720
each colour-difference signal	360	360
7. Analogue-to-digital horizontal timing relationship: from end of digital active line to 0H	16 luminance clock periods (NTSC)	12 luminance clock periods (PAL)
8. Correspondence between video signal levels and quantization levels: scale	0 to 255	(0 to 1023)

(*continued*)

Table 3.1   *(cont.)*

Parameters	525-line, 60 field/s systems	625-line, 50 field/s systems
luminance signal	220 (877) quantization levels with the black level corresponding to level 16 (64) and the peak white level corresponding to level 235 (940). The signal level may occasionally excurse beyond level 235 (940)	
colour-difference signal	225 (897) quantization levels in the centre part of the quantization scale with zero signal corresponding to level 128 (512)	
9. Code-word usage	Code-words corresponding to quantization levels 0 and 255 are used exclusively for synchronization. Levels 1 to 254 are available for video	

Note that the sampling frequencies of 13.5 MHz (luminance) and 6.75 MHz (colour-difference) are integer multiples of 2.25 MHz, the lowest common multiple of the line-frequencies in 525/60 and 625/50 systems, resulting in a static orthogonal sampling pattern for both.

and the colour difference signals are thus sampled to 8- (or 10-) bit depth with the luminance signal sampled twice as often as each chrominance signal (74 ns as against 148 ns). These values are multiplexed together with the structure as follows:

Cb, Y, Cr, Y, Cb, Y, Cr ... etc.

where the three words (Cb, Y, Cr) refer to co-sited luminance and colour difference samples and the following word Y corresponds to a neighbouring luminance only sample. The first video data word of each active line is Cb.

Video timing reference signals (TRS)

The digital active line begins at 264 words from the leading edge of the analogue line synchronization pulse, this time being specified between half amplitude points. This relationship is shown in Figure 3.6. The start of the first digital field is fixed by the position specified for the start of the

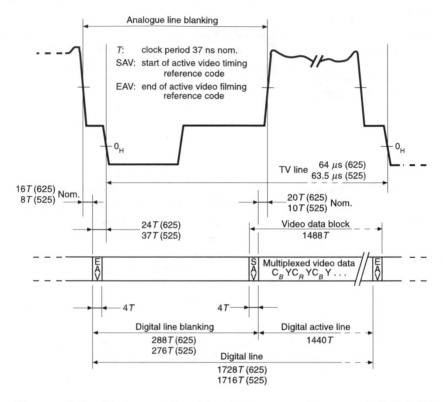

Figure 3.6 *Timing relationships between analogue and digital TV signals*

digital active line; the first digital field starts at 24 words before the start of the analogue line No. 1. The second digital field starts 24 words before the start of analogue line No. 313.

Two video timing reference signals are multiplexed into the data stream on every line, as shown in Figure 3.6, and retain the same format throughout the field blanking interval. Each timing reference signal consists of a four word sequence, the first three words being a fixed preamble and the fourth containing the information defining:

first and second field blanking
state of the field blanking
beginning and end of the line blanking

This sequence of four words can be represented, using hexadecimal notation, in the following manner:

$$FF\ 00\ 00\ XY^{1}$$

in which XY represents a variable word. In binary form this can be represented in the following form:

Data bit no.	First word (FF)	Second word (00)	Third word (00)	Fourth word (XY)
7	1	0	0	1
6	1	0	0	F
5	1	0	0	V
4	1	0	0	H
3	1	0	0	P3
2	1	0	0	P2
1	1	0	0	P1
0	1	0	0	P0

The binary values of F, V and H characterize the three items of information listed earlier:

F = 0 for the first field
V = 1 during the field-blanking interval
H = 1 at the start of the line-blanking interval.

The binary values P0, P1, P2 and P3 depend on the states of F, V and H in accordance with the following table, and are used for error detection/correction of timing data.

F	V	H	P3	P2	P1	P0
0	0	0	0	0	0	0
0	0	1	1	1	0	1
0	1	0	1	0	1	1
0	1	1	0	1	1	0
1	0	0	0	1	1	1
1	0	1	1	0	1	0
1	1	0	1	1	0	0
1	1	1	0	0	0	1

Clock signal

The clock signal is at 27 MHz, there being 1728 clock intervals during each horizontal line period (PAL).

Filter templates

The remainder of CCIR Recommendation 601 is concerned with the definition of the frequency response plots for pre-sampling and reconstruction filter. The filters required by Rec. 601 are practically difficult to achieve, and equipment required to meet this specification has to contain expensive filters in order to obtain the required performance.

Parallel digital interface

The first digital video interface standards were parallel in format. They consisted of 8 or 10 bits of differential data at ECL data levels and a differential clock signal, again as an ECL signal. Carried via a multi-core cable, the signals terminated at either end in a standard D25 plug and socket. In many ways this was an excellent arrangement, and is well suited to connecting two local digital video tape machines together over a short distance. The protocol for the digital video interface is in Table 3.2. Clock transitions are specified to take place in the centre of each data-bit cell.

Table 3.2   *The parallel digital video interface*

Pin no.	Function
1	Clock +
2	System ground
3	Data 7 (MSB) +
4	Data 6 +
5	Data 5 +
6	Data 4 +
7	Data 3 +
8	Data 2 +
9	Data 1 +
10	Data 0 +
11	Data −1 + (10 bit systems only)
12	Data −2 + (10 bit systems only)
13	Cable shield
14	Clock −
15	System ground
16	Data 7 (MSB) −
17	Data 6 −
18	Data 5 −
19	Data 4 −
20	Data 3 −
21	Data 2 −
22	Data 1 −
23	Data 0 −
24	Data −1 − (10 bit systems only)
25	Data −2 − (10 bit systems only)

Problems arose with the parallel digital video interface over medium long distances, resulting in 'mis-clocking' of the input data and visual 'sparkles' or 'zits' on the picture. Furthermore, the parallel interface required expensive and non-standard multi-core cable (although over very short distances it could run over standard ribbon cable), and the D25 plug and socket are very bulky. Today, the parallel interface standard has been entirely superseded by the serial digital video standard, which is designed to be transmitted over relatively long distances using the same coaxial cable as used for analogue video signals. This make its adoption and implementation as simple as possible for existing television facilities converting from analogue to digital video standards.

Serial digital interface

SMPTE 259M specifies the parameters of the serial digital standard. This document specifies that the parallel data in the format given in the previous section be serialized and transmitted at a rate 10 times the parallel clock frequency. For component signals this is:

$$27 \, \text{Mbits/s} \times 10 = 270 \, \text{Mbits/s}$$

The serialized data must have a peak-to-peak amplitude of 800 mV ($\pm 10\%$) across 75 ohms, a nominal rise time of 1 ns and a jitter performance of ± 250 ps. At the receiving end, the signals must be converted back to parallel in order to present the original parallel data to the internal video processing. (Note that no equipment processes video in its serial form although digital routing switchers and DAs, where there is no necessity to alter the signal, only to buffer it or route it, do not decode the serial bitstream.)

Serialization is achieved by means of a system illustrated in Figure 3.7. Parallel data and parallel clock are fed into input latches and thence to a parallel-to-serial conversion circuit. The parallel clock is also fed to a phase-locked-loop which performs parallel clock multiplication (by 10 times). A sync detector looks for TRS information and ensures this is encoded correctly irrespective of 8- or 10-bit resolution. The serial data is fed out of the serializer and into the scrambler and NRZ to NRZI circuit. The scrambler circuit uses a linear feedback shift-register, which is used to pseudo-randomize the incoming serial data. This has the effect of minimizing the DC component of the output serial data stream; the NRZ to NRZI circuit converts long series of ones to a series of transitions. The resulting signal contains enough information at clock rate and is sufficiently DC-free that it may be sent down existing video cables. It may then be re-clocked, decoded and converted back to parallel data at the receiving equipment. Due to its very high data rate, serial video must be carried by ECL circuits. An illustration of a typical ECL gate is given in

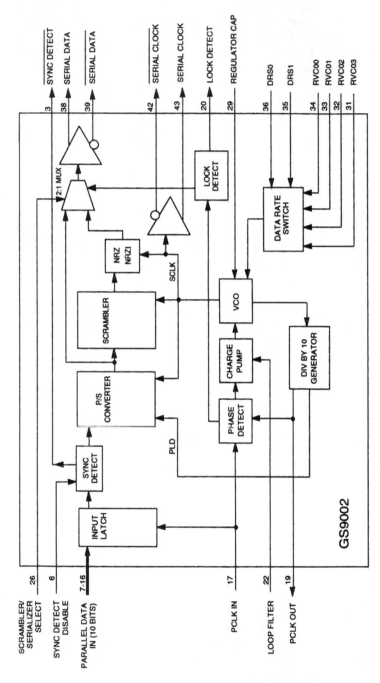

Figure 3.7 Serialization system for bit-serial TV signals

Figure 3.8. Note that standard video levels are commensurate with data levels in ECL logic.

Clearly the implementation of such a high-speed interface is a highly specialized task. Fortunately, practical engineers have all the requirements for interface encoders and decoders designed for them by third-party integrated circuit manufacturers.

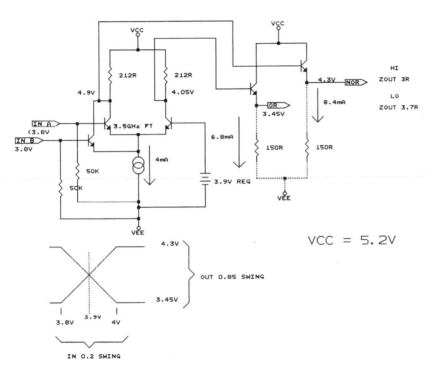

Figure 3.8 *ECL gate schematic*

HDTV serial interface

As already explained, Europe's flirtations with high-definition television (HDTV) have met with disaster. The situation is not the same in the USA. This is probably for a number of reasons, one of which is certainly the comparatively low fidelity of NTSC broadcasts. Coupled with demographic (and geographic) differences between the USA and Europe, this has led to a situation where broadcasters are convinced HDTV will give them a competitive advantage. Clearly equipment manufacturers are eager to provide new equipment for these premium services, and have therefore been enthusiastic in standardizing HDTV coding and interfacing requirements. The standards are SMPTE 240M (coding), SMPTE 260M (parallel interface) and SMPTE 292M (serial interface). The HDTV standard to which these apply is the SMPTE 274M-4; the 1125 line, 60 Hz (or 59.94 Hz) interlaced-field standard with a 16:9 aspect ratio. This provides an image format of 1920 × 1080 active pixels and the coding and interface standards are essentially 'scaled-up' versions of the standard television (SDV) interfaces we have met already. The HDTV parallel interface (SMPTE 274M) is expected to have very limited application; clock speed is 148.5 MHz, balanced ECL, 10 bits and clock. The serial version is specified at 1.485 Gbit/s, and uses 10-bit NRZI coding (just like SMPTE 259M); this can exist on coax or on fibre. Chip-sets have already been developed to support the serial standard.

Digital audio interfaces

As early as the 1970s, manufacturers started to introduce proprietary digital audio interface standards enabling various pieces of digital audio hardware to pass digital audio information directly without recourse to standard analogue connections. Unfortunately, each manufacturer adopted its own standard, and the Sony digital interface (SDIF) and the Mitsubishi interface both bear witness to this early epoch in digital audio technology when compatibility was very poor between different pieces of equipment. However, it wasn't long before customers were demanding an industry-standard interface so that they could mix and match equipment from different manufacturers to suit their own particular requirements. This pressure led to the widespread introduction of standard interfaces for the connection of both consumer and professional digital audio equipment.

The requirements for standardizing a digital interface go beyond those for an analogue interface in that, as well as defining the voltage levels and connector style, it is necessary to define the data format the interface will employ. There are two digital audio interface standards described here:

1. The two-channel, serial, balanced, professional interface (the so-called, AES/EBU or IEC958 type 1 interface)
2. The two-channel, serial, unbalanced, consumer interface (the so-called SPDIF or IEC958 type 2 interface).

In fact both these interfaces are very similar, the variation being more due to electrical differences than to differences in data format.

AES/EBU or IEC958 type 1 interface

This electrically balanced version of the standard digital interface was originally defined in documents produced by the Audio Engineering Society (AES) and the European Broadcasting Union (EBU), and is consequently usually referred to as the AES/EBU standard. This is the standard adopted mainly by professional and broadcast installations. Mechanically this interface employs the ubiquitous XLR connector, and adopts normal convention for female and male versions for inputs and outputs respectively. Electrically, pin 1 is specified as shield and pins 2 and 3 for balanced signal. One of the advantages of the digital audio interface over its analogue predecessor is that polarity is not important, so it is not necessary to specify which pin of 2 and 3 is 'hot'. (Nevertheless, the AES recommends pin 3 is 'phase' or 'hot' and pin 2 is 'return' or 'cold'.) The balanced signal is intended to be carried by balanced, twisted-pair and screen microphone-style cable, and voltage levels are allowed to be between 3 V and 8 V pk–pk (EMF, measured differentially). Both inputs and outputs are specified as transformer coupled and earth-free. The output impedance of the interface is defined as 110 ohms, and a standard input must always terminate in 110 ohms. A drawing for the electrical standard for this interface is given in Figure 3.9.

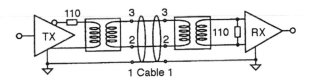

Figure 3.9 *The professional, balanced AES/EBU digital audio interface*

SPDIF or IEC958 type 2 interface

This consumer version of the two-channel serial digital interface is very different electrically from the AES/EBU interface described above. It is a 75-ohm matched termination interface intended for use with coaxial cable. It therefore has more in common with an analogue video signal interface than with any analogue audio counterpart. Mechanically the connector style recommended for the SPDIF interface is RCA style phono, with sockets always being of the isolated type. Voltage levels are defined as 1 V pk–pk when unterminated. Transformer coupling is by no means always used with this interface, but it is recommended at least one end. Figure 3.10 is a drawing of a common implementation of the SPDIF interface.

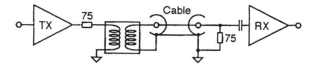

Figure 3.10 *The consumer SPDIF digital audio interface*

Data

Despite the very considerable electrical differences between the AES/EBU interface and the SPDIF interface, their data formats are very similar. Both interfaces have capacity for the real-time communication of 20 bits of stereo audio information at sampling rates between 32 and 48 kHz, as well as provision for extra information which may indicate to the receiving device various important parameters about the data being transferred (such as whether pre-emphasis was used on the original analogue signal prior to digitization). There is also a small overhead for limited error checking and for synchronization. Actually even the 4 'Aux' bits – as shown in Figure 3.11 – have now been pressed into service as audio bits, and the capacity of the AES and SPDIF interface is nowadays considered to be 24-bit.

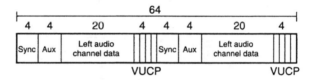

Figure 3.11   *The data-structure of the digital-audio frame*

Some of the earlier digital–audio interfaces, such as Sony's SDIF and the Mitsubishi interface, sent digital audio data and synchronizing data clocks on separate wires. Such standards obviously required multi-core cable and multiway connectors that looked completely different from any analogue interface that had gone before. The intention of the designers of the AES/EBU and SPDIF interfaces was to set standards that created as little 'culture-shock' as possible in both the professional and consumer markets, and they therefore chose connector styles that were both readily available and operationally convenient. This obviously ruled out the use of multi-core and multiway connectors, and resulted in the use of a digital coding scheme that buries the digital synchronizing signals in with the data signal. Such a code is known as 'serial and self-clocking'. The type of code adopted for AES/EBU and SPDIF is bi-phase mark coding. This scheme is sometimes known as Manchester code, and it is the same type of self-clocking serial code used for SMPTE and EBU timecode. Put at its simplest, such a code represents the 'ones and noughts' of a digital signal by two different frequencies where frequency F_n represents a zero and $2F_n$ represents a one. Such a signal eliminates almost all DC content, enabling it to be transformer coupled, and also allows for phase inversion since it is only a frequency (and not its phase) that needs to be detected. The resulting signal has much in common with an analogue FM signal and, since the two frequencies are harmonically related (an octave apart), it is a simple matter to extract the bit-clock from the composite incoming data stream.

In data format terms, the digital audio signal is divided into frames. Each digital audio frame contains a complete digital audio sample for both left and right channel. If 48 kHz sampling is used, it is obvious that 48 000 frames pass over the link in every second, leading to a final baud rate of 3.072 Mbit/s. If 44.1 kHz sampling is employed, 44 100 frames are transmitted every second, leading to a final baud rate of 2.8224 Mbit/s. The lowest allowable transfer-rate is 2.084 Mbit/s, when 32 kHz is used. Just as each complete frame contains a left and right channel sample, so each frame may be further divided into individual audio samples known as sub-frames. A diagram of a complete frame consisting of two sub-frames is given in Figure 3.11.

It is manifestly extremely important that any piece of equipment receiving the digital audio signal as shown in Figure 3.11 must know where the boundaries between frames and sub-frames lie. That is the purpose of the 'sync preamble' section of each frame and sub-frame. The sync preamble section of the digital audio signal differs from all the other data sent over the digital interface in that it violates the rules of a bi-phase mark encoded signal. In terms of the FM analogy given above, you can think of the sync preamble as containing a third non-harmonically related frequency which, when detected, establishes the start of each sub-frame. There exists a family of three slightly different sync preambles. One marks the beginning of a left sample sub-frame and another marks the start of the right channel sub-frame; the third sync-preamble pattern is used only once every 192 frames (or once every 4 milliseconds in the case of 48 kHz sampling), and is used to establish a 192-bit repeating pattern to the channel-status bit labelled C in Figure 3.11.

The 192-bit repeat pattern of the C bit builds up into a table of 24 bytes of channel-status information for the transmitted signal. It is in this one bit of data every sub-frame that the difference between the AES/EBU interface data format and the SPDIF data format is at its most significant. The channel status bits in both the AES/EBU format and SPDIF format communicate to the receiving device such important parameters as sample-rate, whether frequency pre-emphasis was used on the recording etc. Channel status data is normally the most troublesome aspect of practical interfacing using the SPDIF and AES/EBU interface – especially where users attempt to mix the two interface standards. This is because the usage of channel status in consumer and professional equipment is almost entirely different. It must be understood that the AES/EBU interface and the SPDIF interface are thus strictly incompatible in data-format terms, and the only correct way to transfer data from SPDIF to AES/EBU and

AES/EBU to SPDIF is through a properly designed format converter, which will decode and re-code the digital audio data to the appropriate standard.

Other features of the data format remain pretty constant across the two interface standards. The validity bit, labelled V in Figure 3.11, is set to 0 every sub-frame if the signal over the link is suitable for conversion to an analogue signal. The user bit, labelled U in Figure 3.11, has a multiplicity of uses defined by particular users and manufacturers. It is most often used over the domestic SPDIF interface. The parity bit, labelled P in Figure 3.11, is set such that the number of ones in a sub-frame is always even. It may be used to detect individual bit errors but not conceal them. It's important to point out that both the AES/EBU interface and its SPDIF brother are designed to be used in an error-free environment. Errors are not expected over digital audio links, and there is no way of correcting for them.

Practical digital audio interface

There are many ways of constructing a digital audio interface, and variations abound from different manufacturers. Probably the simplest consists of an HC-family inverter IC, biased at its midpoint with a feedback resistor and protected with diodes across the input to prevent damage from static or over-voltage conditions. (About the only real merit of this circuit is simplicity!) Transformer coupling is infinitely preferred. Happily, whilst analogue audio transformers are complex and expensive items, digital audio – containing as it does no DC component and very little low-frequency component – can be coupled via transformers which are tiny and cheap! So, it represents a false economy indeed to omit them in the design of digital interfaces. Data-bus isolators manufactured by Newport are very suitable. Two or four transformers are contained within one IC-style package. Each transformer costs about £1.50 – a long way from the £15 or so required for analogue transformers. Remember, too, that 'in digits' only one transformer is required to couple both channels of the stereo signal. You'll notice, looking at the circuit diagram (Figure 3.12), that RS422 (RS485) receiver-chips buffer and re-slice the digital audio data. The SN75173J is a quad receiver in a single 16-pin package costing a few pounds. The part has the added advantage that, to adapt the interface between SPDIF and AES, all that is required is to change the value of the terminating resistor on the secondary side of the input transformer. Digital output driving is performed by an RS422 driver IC.

TOSlink optical interface

In many ways an optical link is the ideal solution for joining two pieces of digital audio equipment together. Obviously, a link that has no electrical contact cannot introduce ground-loop hum problems. Also, because the bandwidth of an optical link is so high, it would appear from a superficial inspection that an optical link would provide the very fastest (and therefore 'cleanest') signal path possible. However, the speed of cheap commercial digital links is compromised by the relatively slow light-emitting diode transmitters and photo-transistor receivers housed within the connector shells and cheap optical fibres, which allow the optical signal more than one direct path between transmitter and receiver (the technical term is multimodes). These cause a temporal smearing of the audio pulses, resulting in an effect known as modal dispersion. This can cause a degree of timing instability in digital audio circuits (jitter), and this can affect sound quality. The only advantage the optical link confers, therefore, is its inherent freedom from ground-path induced interference signals such as hum and RF noise. Yet at digital audio frequencies, ground isolation – if it is required – is much better obtained by means of a transformer. Another disadvantage is the cost of optical links, which are, metre for metre, 10–25 times as expensive as a coaxial interface. Unfortunately, many pieces of commercial equipment only feature SPDIF interfaces on optical connectors. If you want to modify a piece of equipment with an optical interface to include SPDIF coaxial output, a modification is shown in Figure 3.13.

Unbalanced (75 ohm) AES interface

In October 1995, the AES produced an information document (*AES-3id-1995*) relating to the transmission of digital audio information (utilizing the professional data format) over an electrical interface that has much in common with the interconnection standards employed in analogue video. Limitations of AES data travelling on twisted pairs and terminated in XLRs include poor RF radiation performance and a limitation of maximum transmission distance to 100 metres. The proposed unbalanced interface is suitable for transmission distances of up to 1000 metres. Furthermore, by a prudent choice of impedance and voltage operating level coupled with a sensible specification of minimum rise-time, the signal is suitable for routing through existing analogue video cables, switchers and distribution amplifiers.

The salient parts of the signal and interface specification are given in Table 3.3.

This interface standard is becoming more and more prevalent in the world of professional television broadcast for the handling of digital audio signals, and is fast replacing the balanced 110-ohm interface.

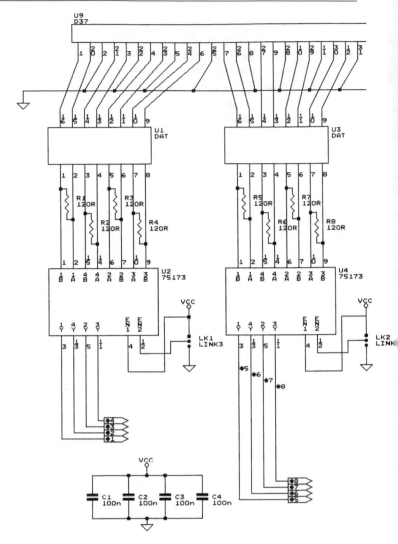

Figure 3.12(a)   *Practical digital audio input interfaces*

Serial multi-channel audio digital interface (MADI)

The MADI standard is a serial transmission format for multi-channel linearly represented PCM audio data. The specification covers transmission of 56 mono 24-bit resolution channels of audio data with a common sampling frequency in the range of 32–48 kHz. Perhaps this is more easily conceived of in terms of 28 stereo 'AES' audio channels (i.e. of AES3-1985 data) travelling on a common bearer, as illustrated in Figure 3.14. The

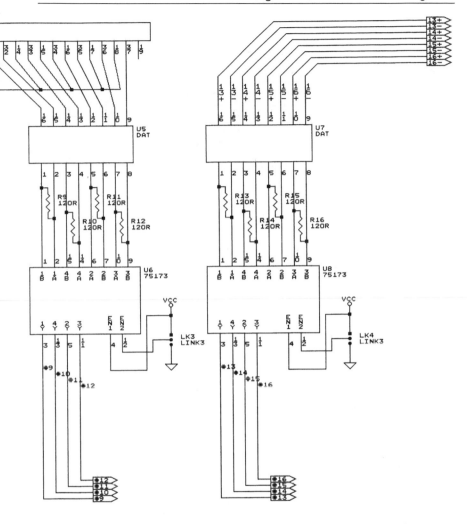

MADI standard is not a 'networking' standard; in other words, it only supports point-to-point interconnections.

Data format

The MADI serial data stream is organized into frames which consist of 56 channels (numbered 0–55). These channels are consecutive within the

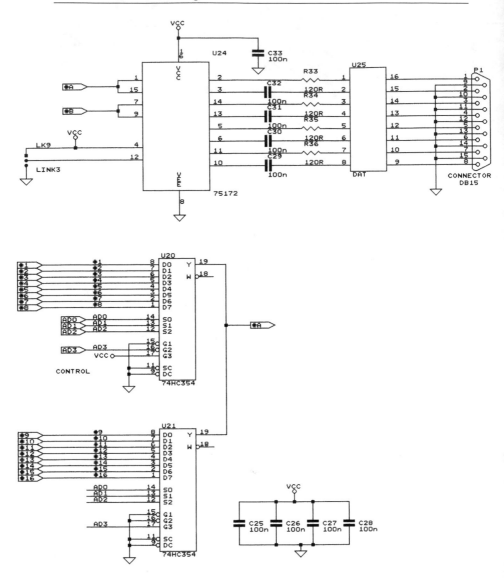

Figure 3.12(b)   *Digital audio switching and output interfaces*

frame and the audio data remains, just as it is in the original digital audio interface, in linearly coded 2s-complement form – although this is scrambled as described below. The frame format is illustrated in Figure 3.14. Each channel 'packet' consists of 32 bits, of which 24 are allocated to audio data (or possibly non-audio data if the non-valid flag is invoked) and 4 bits for the validity (V), user (U), channel-status (C) and parity (P)

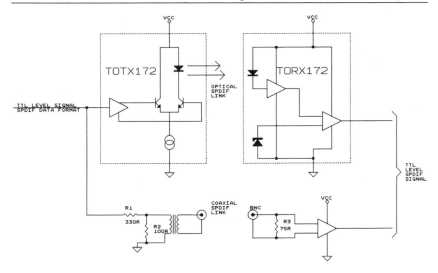

Figure 3.13 *Modification to TOSlink interface to provide SPDIF output*

Table 3.3 *General and signal specification of unbalanced AES interface*

General	
Transmission data format	Electrically equivalent to AES
Impedance	75 ohms
Mechanical	BNC connector

Signal characteristics	
Output voltage	1 V, measured when terminated in 75 ohms
DC offset	less than 50 mV
Rise/fall time	30–44 ns
Bit width (at 48 kHz)	162.8 ns

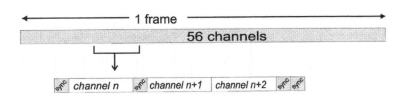

Figure 3.14 *Data structure of MADI, multi-channel audio interface*

bits as they are used in the AES3-1985 standard audio interface. In this manner the structure and data within contributing dual-channel AES bitstreams can be preserved intact when travelling in the MADI multi-channel bitstream. The remaining 4 bits per channel (called, confusingly, mode-bits) are used for frame synchronization on the MADI interface and for preserving information concerning A/B pre-ambles and start of channel-status block within each of the contributing audio channels.

Scrambling and synchronization

Serial data is transmitted over the MADI link in polarity-insensitive (NRZI) form. However, before the data is sent it is subjected to a 4-bit to 5-bit encoding as defined in Table 3.4.

MADI has a rather unusual synchronization scheme in order to keep transmitter and receiver in step. The standard specifies that the transmitter inserts a special synchronizing sequence (1100010001) at least once per frame. Note that this sequence cannot be derived from data, as specified in Table 3.4. Unusually, this sync signal need not appear between every channel 'packet', and is simply repeated wherever required in order to regulate the final data rate of 100 Mbits/s specified in the standard.

Table 3.4   *Data encoding prior to transmission over the MADI link*

Input data sequence	Output data sequence
0000	11110
0001	01001
0010	10100
0011	10101
0100	01010
0101	01011
0110	01110
0111	01111
1000	10010
1001	10011
1010	10110
1011	10111
1100	11010
1101	11011
1110	11100
1111	11101

Electrical format

MADI travels on a coaxial cable interface with a characteristic impedance of 75 ohms. Video-style BNC connectors are specified. Because the signal output is practically DC-free, it may be AC coupled and must sit around $0\,V \pm 100\,mV$. This signal is specified to have a peak-to-peak amplitude of 300–600 mV when terminated – this choice of amplitude being determined by the practical consideration that the signal could be directly derived from the output of an ECL gate (see discussion of serial digital video format).

Fibre optic format

Oddly, the MADI standard did not define a fibre implementation, despite the fact that the copper implementation was based on a widely used fibre interface known as FDDI (ISO 9314). It is this standard, which pre-dates the MADI, which specified the 4-bit to 5-bit mapping defined in Table 3.4! This lack of standardization has resulted in a rather disorganized situation regarding MADI over fibre. The AES's own admission is simply that 'any fibre-system could be used for MADI as long as the basic bandwidth and data-rate can be supported ... However, adoption of a common implementation would be advantageous.'

Embedded audio in video interface

So far, we have considered the interfacing of digital audio and video separately. Manifestly, there exist many good operational reasons to combine a television picture with its accompanying sound 'down the same wire'. The standard that specifies the embedding of digital audio data, auxiliary data and associated control information into the ancillary data space of the serial digital interconnect conforming to SMPTE 259M in this manner is the proposed SMPTE 272M standard.

The video standard has adequate 'space' for the mapping of a minimum of one stereo digital audio signal (or two mono channels) to a maximum of eight pairs of stereo digital audio signals (or 16 mono channels). The 16 channels are divided into four audio signals in four 'groups'. The standard provides for 10 levels of operation (suffixed A to J) which allow for various different and extended operations over and above the default synchronous 48 kHz/20 bit standard. The audio may appear in any and/or all the line blanking periods, and should be distributed evenly throughout the field. Consider the case of one 48 kHz audio signal multiplexed into a 625/50 digital video signal. The number of samples to be transmitted every line is

$$(48000)/(15625)$$

which is equivalent to 3.072 samples per line. The sensible approach is taken within the standard of transmitting three samples per line most of the time and transmitting four samples per line occasionally in order to create this non-integer average data rate. In the case of 625/50, this leads to 1920 samples per complete frame. (Obviously a comparable calculation can be made for other sampling and frame rates.) All that is required to achieve this 'packeting' of audio within each video line is a small amount of buffering either end and a small data overhead to 'tell' the receiver whether it should expect three or four samples on any given line.

Figure 3.15 illustrates the structure of each digital audio packet as it appears on preferably all, or nearly all, the lines of the field. The packet starts immediately after the TRS word for EAV (end of active line) with the ancillary data header 000,3FF,3FF. This is followed by a unique ancillary data ID that defines which audio group is being transmitted. This is followed with a data-block number byte. This is a free-running counter counting from 1 to 255 on the lowest 8 bits. If this is set to zero, a de-embedder is to assume this option is not active. The 9th bit is even parity for b7 to b0, and the tenth is the inverse of the ninth. It is by means of this data-block number word that a vertical interval switch could be discovered and concealed. The next word is a data count, which indicates to a receiver the number of audio data words to follow. Audio sub-frames then follow as adjacent sets of three contiguous words. The format in which each AES subframe is encoded is illustrated in Figure 3.16. Each audio data-packet terminates in a checksum word.

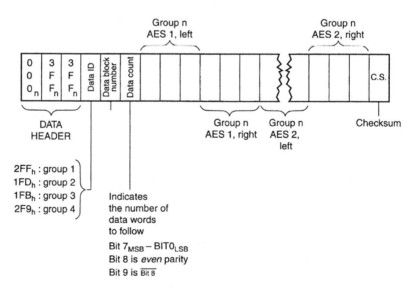

Figure 3.15 *Data format for digital audio packets in the SDV bitstream*

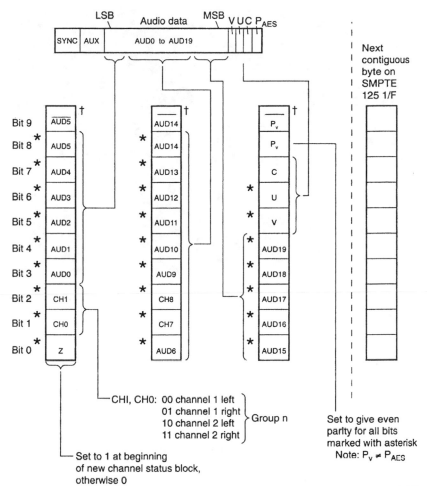

Figure 3-16 *Each AES subframe is encoded as three contiguous data packets*

The standard also specifies an optional audio control packet. If the control packet is not transmitted, a receiver defaults to 48 kHz synchronous operation. For other levels, the control byte must be transmitted in field interval. Embedded audio is viewable using a picture monitor with a 'pulse-cross' option, as illustrated in Figure 3.17.

The monitor is fed with delayed line and field sync signals derived from the original TRS information. Audio data is viewable in the line-blanking interval as moving 'dots'. This is a handy way of checking if a signal has embedded audio or not.

Figure 3.17 *Embedded digital audio as viewed on pulse-cross monitor*

Error detection and handling

Error detection and handling (EDH) is a system devised to assess the quality of serial digital video links (with and without embedded digital audio). The necessity for EDH arises because of the nature of digital transmission links. As we have seen, digital signals have a great advantage over their analogue forebears in that they can suffer a great deal more distortion before artefacts or noise become visible or audible. Digital signals are said to be more 'robust'.

Unfortunately, this characteristic has one unfortunate consequence; that a serial digital link can be very close to the point of complete failure yet still providing excellent results. This trait is described as the 'cliff effect', the term aptly relating the phenomenon that a transmission link will deliver perfect results in the presence of distortion and noise up until some definable point at which the deterioration of the received signal will be catastrophic. An analogue signal, on the other hand, betrays a gradual deterioration of signal quality in the presence of various forms of distortion or noise; it is said to have a much 'softer' (or more gradual) failure mechanism (see Figure 3.18).

More importantly – and here think back to the hieroglyph analogy drawn earlier in this book – an analogue signals betrays the distortion done to it; the distortion and the signal are inseparable. So, measuring the video or audio signal, will reveal the distortion 'imprinted' onto it. A ditigal

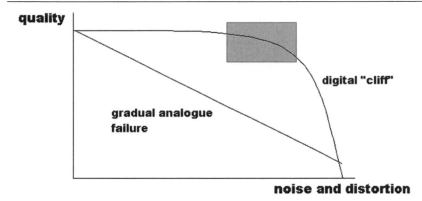

Figure 3.18 *The digital 'cliff effect', where increasing errors fail to give any warning of trouble until the error rate is so high that the signal falls completely*

signal, being – by contrast – 'iconic' rather than representational, does not betray the distortion it encounters on its signal path. That's not to say that the basic physical reality of noise and distortion are of no consequence. These physical effects will certainly have an effect by eventually rendering a digital '1' into a digital '0' or vice-versa; a phenomenon known as a digital 'error'. But, provided the amount of errors within a given time frame (known as 'error rate') are within the error correction and concealment capabilities of the system, you simply won't hear or see the result of transmission medium noise or distortion on a recovered digital audio or video signal. Up until the point at which sufficient damage is done to the signal that the receiving mechanism can no longer 'cope' with the error rate, whereupon the signal will become useless.

This has important operational ramifications, because it can mean that digital television studios and network engineers can be operating very near the point of complete failure (shown by the grey box in Figure 3.18) without knowing it! So, because the digital signal itself – unlike its analogue cousin – does not reveal the distortion (or error rate) it experiences, it's necessary to include extra information along with digital signal, to indicate to the receiver the degree of distortion (or number of errors) resulting from its transmission through the transmission medium. This extra information is known, in serial digital video circles, as EDH.

EDH codeword generation

We will look at data protection using parity generation in Chapter 9. An alternative approach, often used in computing, is to send a checksum value; a numerical value which is some form of binary addition of the number of bytes (or bits) contained in a file. Subsequent recalculation of this value, and comparison with the original checksum, quickly reveals any errors which have befallen the original data. This is a fine idea, but where lots of data is involved (as is the case with digital video) a straightforward addition would result in an impractically large number. Instead a method is used to generate a practically unique, short codeword from a long train of input data. If this codeword generation is performed at both the sending end – where it is sent along with the data – and at the receiving end and the encoded codeword is afterwards compared with the locally generated codeword, any disparity clearly indicates the presence of errors in the transmission path.

In EDH for digital video, the check-word is calculated using the CRC (Cyclic Redundancy Code) polynomial generation method which is illustrated in Figure 3.19; this function is identical in both receiver and transmitter. EDH information is calculated for each video field and is sent as an ancillary signal in spare data-space in the following video field; in fact on line 5 of field 1 and line 318 of field 2 in 625/50 systems. (Note that this position ensures the EDH data is before the standardized switching point, so that errors are not generated when switching between different bitstreams.)

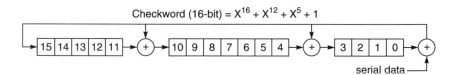

$$\text{Checkword (16-bit)} = X^{16} + X^{12} + X^5 + 1$$

Figure 3.19   *CRC generation mechanism*

In a simple world, one EDH codeword would be sufficient, calculated at – for instance – the camera and checked at the master-control room output. But real television studios do not represent a simple, linear signal path and most units within the signal-path, with the exception of routing switchers and distribuiton equipment, are in-circuit in order to affect picture changes. Video mixing, keying, colour correction and so on, all change picture content and would thereby invalidate the EDH codeword value calculated at the original source. There is therefore a requirement to decode and recode EDH values within each piece of TV studio equipment. In addition there are many pieces of television equipment (and especially those involving signal storage) where the picture may be passed without signal processing, but where the input-signal blanking-intervals (field and line) are simply discarded and regenerated on the output of the equipment. For this reason EDH codewords are calculated both for the active-picture (AP) portion of the SDI signal and for the entire picture field; or full-field (FF). Note that this means that a signal error occurring in picture time will corrupt both the codeword for AP and for FF but an error in the blanking signal period will corrupt the CRC of the FF but not the codeword for AP. A further checksum is also calculated for the ancillary data within the SDI bitstream, this is known as the ANC.

EDH flags

As explained, EDH was invented so that errors in digital links could be detected and remedied, thereby ensuring a well-maintained digital TV studio. In order to report on the status of digital links and apparatus, any digital TV equipment incorporating EDH detection and generation circuitry has a standardized 'reporting system', known as EDH status flags. These flags – each a single digital bit – are inserted back into the outgoing digital video signal and may also perform a number of functions. These may range from the simplest, like lighting a series of LEDs, to reporting these 'flags' back to a centralized status and error-recording computer. This computer may display the entire studio (or network) signal flow in diagrammatic form, highlighting parts of the signal circuit with errors and initiating various operator alarms. Indeed this form of highly organized 'status monitoring' function – so important in the reality of today's world of low-manning – is one of the chief driving forces behind the adoption of EDH capable equipment.

Each of the three EDH codewords has a series of five flags; resulting in a possible fifteen flags:

	EDH	EDA	IDH	IDA	UES
AP	×	×	×	×	×
FF	×	×	×	×	×
ANC	×	×	×	×	×

Each of the flags are defined thus:

EDH   Error detected here
EDA   Error detected already (i.e. upstream)
IDH   Internal error detected here (i.e. occurring within a piece of equipment)
IDA   Internal error detected already (i.e. upstream)
UES   Unknown error status

If an EDH-equipped device receives a clean-signal with no digital errors, none of the 'here' flags is set. If no EDH information is encoded on the incoming signal, then the UES (unknown error status) flag is set and new codewords are generated for the outgoing signal. In a studio where all links are working properly, none of the flags will normally be set except perhaps the UES – due to some equipment not being EDH-equipped.

Now imagine there is an error between equipment B and downstream equipment C, connected by cable (B–C). Equipment C will raise the EDH (error detected here) flag, showing that link (B–C) is under stress and not operating correctly. It will pass this flag to the next piece of equipment in the chain (D), which – provided it too does not detect errors in link (C–D) – will raise the EDA (error detected already) flag and reset the EDH flag. An operator looking at a schematic representation of the studio like this:

A ————— B ————— C ————— D ————— E
(no flags)   (no flags)   (EDH)   (EDA)   (EDA)

will be able to tell, by straightforward inspection of the EDH and EDA flags, where a fault has occurred within the system and which part of the signal-circuit is at fault. The enduring presence of the EDA flag as the signal is passed down the signal chain, also warns the operator that the signal includes errors and is possibly unsuitable for transmission.

IDH and IDA flags extend this idea to the detection and generation of flags which indicate if errors have occurred inside equipment.

Reference

Vanderkooy, J. and Lipshitz, S. (1984). Resolution below the least significant bit in digital systems with dither. *J. Audio. Eng. Soc.* **32**, 106–113.

Note

1 TRS words are defined here in their 8-bit form. In the 10-bit version, the preamble is 3FF, 000, 000, XYZ. Because of the existence of both 8-bit and 10-bit equipment, for detection purposes, all values in the region 000 to 003 and 3FC to 3FF must be considered equivalent to 000 and 3FF respectively.

4
Digital signal processing

Introduction

Digital signal processing (DSP) involves the manipulation of real-world signals (for instance audio signals, video signals, medical or geophysical data signals etc.) within a digital computer. Why might we want to do this? Because these signals, once converted into digital form (by means of an analogue-to-digital converter – see Chapter 3), may be manipulated using mathematical techniques to enhance, change or display the data in a particular way. For instance, the computer might use height or depth data from a geophysical survey to produce a coloured contour map, or the computer might use a series of two-dimensional medical images to build up a three-dimensional virtual visualization of diseased tissue or bone. Another application, this time an audio one, might be to remove noise from a music signal by carefully measuring the spectrum of the interfering noise signal during a moment of silence (for instance, during the run-in groove of a record) and then subtracting this spectrum from the entire signal, thereby removing only the noise – and not the music – from a noisy recording.

DSP systems have been in existence for many years but, in these older systems, the computer might take many times longer than the duration of the signal acquisition time to process the information. For instance, in the case of the noise reduction example, it might take many hours to process a short musical track. This leads to an important distinction that must be made in the design, specification and understanding of DSP systems; in non-real-time systems the processing time exceeds the acquisition or presentation time, whereas real-time systems complete all the required mathematical operations so fast that the observer is unaware of any delay in the process. When we talk about DSP, it's always important to distinguish between real-time and non-real-time DSP.

Digital manipulation

So, what kind of digital manipulations might we expect? Let's think of the functions that might be performed within, for example, a digital sound mixer. First, there is addition. Clearly, at a fundamental level, that is what an audio mixer is – an 'adder' of signals. Secondly, we know that we want to be able to control the gain of each signal before it is mixed. So multiplication must be needed too. So far, the performance of the digital signal processing 'block' is analogous with that of its analogue counter-parts. The simplest form of digital audio mixer is illustrated in Figure 4.1. In this case, two digital audio signals are each multiplied by coefficients (k_1 and k_2) derived from the position of a pair of fader controls; one fader assigned to either signal. The signals issuing from these multiplication stages are subsequently added together in a summing stage. All audio mixers possess this essential architecture, although it may be supplement-ed many times over.

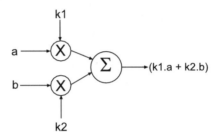

Figure 4.1   *Simple digital mixer*

In fact the two functions of addition and multiplication, plus the ability to delay signals easily within digital systems, allow us to perform all the functions required within a digital sound mixer; even the equalization functions. That's because equalization is a form of signal filtering on successive audio samples, which is simply another form of mathematical manipulation – even though it not usually regarded as such in analogue circuitry.

Digital filtering

The simplest form of analogue low-pass filter is shown in Figure 4.2a. Its effect on a fast rise-time signal wavefront (an 'edge') is also illustrated. Notice that the resulting signal has its edges slowed-down in relation to the incoming signal. Its frequency response is also illustrated, with its turnover frequency at $f3$. In digital circuits there are no such things as capacitor or inductors, which may be used to change the frequency response of a circuit. Instead, digital filters work by the interaction of

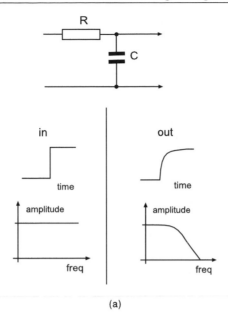

(a)

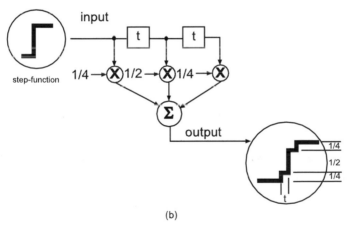

(b)

Figure 4.2   (a) *Action of a simple analogue (RC) filter; (b) Action of a digital FIR filter – compare result with (a)*

signals delayed with respect to one another. This principle is the basis behind all digital filtering, and may be extended to include several stages of delay as shown in Figure 4.2b. By utilizing a combination of adder and variable multiplication factors (between the addition function and the signal taps), it's possible to achieve a very flexible method of signal filtering in which the shape of the filter curve may be varied over a very wide range of shapes and characteristics. Whilst such a technique is possible in analogue circuitry, note that the 'circuit' (shown in Figure 4.2b) is actually not a real circuit at all, but a notional block diagram. It is in the realm of digital signal processing that such a filtering technique really comes into its own: The DSP programmer has only to translate these processes into microprocessor type code to be run on a micro-controller IC that is specifically designed for audio or video applications – a so-called DSP IC. Herein lies the greatest benefit of digital signal processing; that, by simply re-programming the coefficients in the multiplier stages, a completely different filter may be obtained. Not only that, but if this is done in real-time too, the filter can be made adaptive; adjusting to demands of the particular moment in a manner which might be useful for signal compression or noise reduction. Digital filtering is treated in greater detail at the end of this chapter.

Digital image processing

The real advantage of treating images in the digital domain is the ability it affords to filter, manipulate and distort the image. Here we examine how image manipulations are implemented. The simplest form of digital image processing filter is a point operation type. The point operation filter usually has a cousin in the analogue world. Some of the most common filters are discussed below.

Point operations

1 *Contrast, luminance-gain or Y gain.* This controls the gain of all three primary channels simultaneously; it therefore has no effect upon colour information. Viewed in terms of the colour three-space illustrated in Figure 2.3 in Chapter 2, Y gain has the effect of multiplying pixel values and of stretching or contracting colour three-space along the achromatic, luminance dimension.
2 *Brightness, luminance black level, Y black level.* This control allows a 'bias' or offset value to be applied to all values of luminance pixels. This has the effect of creating a 'lift' or 'sink' over the whole picture. Because incorrect DC (or black) level is so often a problem with real pictures (perhaps because of incorrect lens aperture), provision is usually made for the control to work in a positive and negative sense. For instance, if the picture appears too light when unprocessed, the operator need simply adjust the Y DC level control negatively to counteract the elevated values in the picture.
3 *U gain and V gain.* U and V gain controls act solely upon the chrominance value of the image pixels. U gain has the effect of increasing or decreasing the chrominance values along the blue–

yellow colour axis; V gain has the effect of increasing or decreasing the chrominance values along the red/magenta–green/cyan axis. These U and V axes are the colour axes used in television coding and decoding, and are subsequently liable to various distortions in this equipment. It is these scalar distortions that these controls are intended to correct.

4 *Red, green and blue black level.* Red, green and blue black level controls allow a 'bias' or offset value to be applied to all pixels of the selected colour. This has the effect of creating a 'wash' of colour over the whole picture. Because such a wash or colour-caste is so often a problem with real pictures (perhaps because of incorrect colour-balance), provision is usually made for the controls to work in a positive and negative sense. For instance, if the picture appears too red when unprocessed, the operator needs simply to adjust the red black level control negatively to counteract the elevated red values in the picture.

5 *Red, green and blue contrast or gain.* Red, green and blue contrast controls allow individual positive and negative adjustment of the contrast range (or intensity) of each of the three primary colours. These controls are especially useful for the control of the highlights in a picture. Sometimes highlights can appear very yellow on some tape and film formats. Increasing the blue gain will often restore a more natural colorimetry.

6 *Black paint.* A black-level paint control allows control of the intensity of colour only in the black areas of the picture. Often film footage displays this effect, where colorimetry may be good outside the black region but the black may, for instance, take on a red tinge.

7 *Gamma.* Red, green and blue gamma controls adjust the linearity of the transfer characteristic of each primary colour. These controls may be used to adjust the colour-balance in the grey areas of the incoming picture. Adjustment of these controls leaves the colour-balance of black and white levels unaltered.

8 *White paint.* Red, green and blue white paint controls allow individual positive and negative adjustment of the colorimetry of the higher greys and whites of the picture. This control is especially useful for the colour-correction of picture highlights. Sometimes highlights can appear very yellow on some tape and film formats. Increasing the blue white paint will often restore a more natural colorimetry.

Window operations

Window operations are more complex than point operations and involve the computation of individual pixel values based upon the values of pixels in the neighbourhood. The general form of a three by three window filter is illustrated in Figure 4.3. The process being performed in computing the

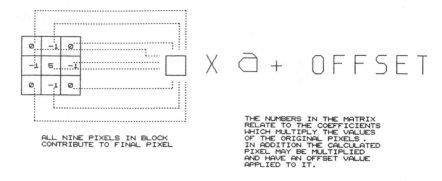

ALL NINE PIXELS IN BLOCK
CONTRIBUTE TO FINAL PIXEL

THE NUMBERS IN THE MATRIX
RELATE TO THE COEFFICIENTS
WHICH MULTIPLY THE VALUES
OF THE ORIGINAL PIXELS .
IN ADDITION THE CALCULATED
PIXEL MAY BE MULTIPLIED
AND HAVE AN OFFSET VALUE
APPLIED TO IT.

Figure 4.3   *General form of 3 × 3 convolutional window-operation filter*

new pixel value from the others is known as convolution. Some common examples of window operation filters are given below. The commonest type is known in television as an aperture filter. This control governs the high spatial frequency response. This control can be used either to 'crispen' a picture or to reduce HF performance to decrease the deleterious effects of video noise. In order to see how the filter achieves its effect, consider the synthetic television image in Figure 4.4. We are looking very closely at our image so that we can see individual pixels.

The aperture filters has the following convolution mask:

$$\begin{array}{ccc} 0 & -1 & 0 \\ -1 & 5 & -1 \\ 0 & -1 & 0 \end{array}$$

(Constant operator: divide by 1; bias: 0.)

Look at Figure 4.5 (which shows the effect of this convolution mask upon the image in Figure 4.4); notice how the edges of the grey to white transition have been magnified by the action of the sharpening filter. Now let's take a second synthetic image (the negative of the first) and look at the

Figure 4.4   *A synthetic TV image*

Figure 4.5   *Effect of the sharpening filter on Figure 4.4*

action of a low-pass blur filter. In many ways the blur filter is the easiest to understand; all pixels in the neighbourhood contribute to the computed pixel. (Note the examples are derived from computations using three by three matrices, but there is nothing fundamental about this – indeed, results get better and better as a function of window size, within practical limits.) You can see the blur filter causing image components to 'bleed' into associated areas, reducing detail. The blur convolution mask looks like this:

$$\begin{array}{ccc} 1 & 1 & 1 \\ 1 & 1 & 1 \\ 1 & 1 & 1 \end{array}$$

(Constant operator: divide by 9; bias: 0)

The effect of a blur filter is illustrated in Figure 4.6. Notice that the smallest single pixel 'fleck' has nearly disappeared in this blurred image. You might think that this would be very useful in 'cleaning-up' a noisy or grainy picture, if it could be achieved as an effect in isolation, and indeed it is. Noise reduction is a common application of digital image processing. However, the low-pass filter has a deleterious effect on image quality and is therefore unsuitable. What is required is a filter that can remove noise whilst preserving image detail. Fortunately, just such a filter exists; it is known as a median filter. In the median filter, the final pixel is computed as the median value of all pixels in the 5-pixel cruciform-shaped window. Its action is therefore biased against non-correlated information and in favour of correlated information. Figure 4.7 illustrates the action of the median filter. Notice how the individual pixel disappears as a result of this filter. This is ideal because noise – by definition random – does not correlate with adjacent pixels, whereas picture information, which is correlated with adjacent pixels, remains relatively unmolested.

Figure 4.8 demonstrates the effect of the median filter on random noise in the context of a real image.

Figure 4.6 *Effect of a blur filter on the negative version of Figure 4.4*

Figure 4.7 *Effect of the median filter*

(a)   (b)

Figure 4.8   (a) *Noisy TV image;* (b) *Effect of convolution with the median filter on image (a)*

Another useful image manipulation tool is an edge-detect filter. The edge detection mask looks like this:

$$\begin{array}{ccc} 1 & -1 & -1 \\ -1 & 8 & -1 \\ -1 & -1 & -1 \end{array}$$

(Constant operator: divide by 1; bias: 0)

Its effect on our synthetic image is shown in Figure 4.9.

Finally, take a look at the effect of a convolution that creates an embossed effect from a graphic image (Figure 4.10). When used on a photographic image, this creates a very dramatic effect.

Figure 4.9   *Effect of edge-detection filter*

Figure 4.10   *Effect of emboss filter*

The emboss convolution mask looks like this:

$$
\begin{matrix}
-1 & 0 & 0 \\
0 & 0 & 0 \\
0 & 0 & -1
\end{matrix}
$$

(Constant operator: divide by 1; bias: $+127$)

Transforming between time and frequency domains

In considering the digital audio filter above, we noted that adaptation of the frequency response was achieved by the interaction of signals delayed with respect to one another. In other words, a frequency domain manipulation was achieved by treating the signal in the time domain. So important is the ability to 'translate' between the world of frequency and time that the following section concentrates on the techniques needed for so doing. In image processing, the ideas are a bit different. Here, instead of transforming between the time and frequency domain, we are transforming between the spatial intensity and spatial frequency domains. These techniques are explained below too.

Most descriptions of the techniques of transforming between the time and frequency domains commence with an apology that any description is necessarily mathematical. This one, unfortunately, is no different. However, many authors state this is necessary because a rational description of the process is simply impossible. If, like me, even an understanding of the mechanical processes of mathematics still leaves something of a 'hole in the stomach' because I haven't really understood the physical principles, this is a depressing pronouncement. Furthermore, I believe it's unnecessarily gloomy. In the following I have tried to give a description of the continuous Fourier transform and the Discrete Fourier transform in their 1D and 2D forms that is both full and yet intuitive as well. I hope expert mathematicians will forgive some of the simplifications that are inevitable in such an approach.

Fourier transform

In many practical fields, signals are more often thought of in the time domain (or time base) rather than in the frequency domain (or frequency base) – sound signals and television signals are both good examples. But how do we get from the time-domain description (a sine function for example) to a frequency-domain description? In fact the process is very simple, if a little labour intensive. The easiest way to imagine the process is to consider the way a spectrum analyser works.

Consider the simplest input waveform of all – a single pure sine wave. When this signal is input to a spectrum analyser, it's multiplied within the unit by another variable frequency sine wave signal. This second signal is internally generated within the spectrum analyser, and is known as the *basis function*. As a result of the multiplication, new waveforms are generated. Some of these waveforms are illustrated in Figure 4.11.

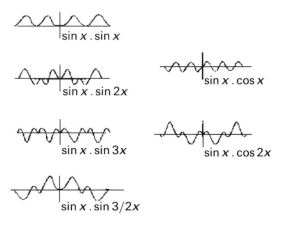

sin *x* . sin *x*

sin *x* . sin 2*x*

sin *x* . cos *x*

sin *x* . sin 3*x*

sin *x* . cos 2*x*

sin *x* . sin 3/2*x*

Figure 4.11   *Multiplications of various functions*

The resulting signals are subsequently low-pass filtered (note that this is the same thing as saying the time-integral is calculated), and the resulting steady voltage is used to drive some form of display device. The principle involved may be appreciated without recourse to difficult mathematics, but one axiom must be understood: When two sine waves are multiplied together, the result contains only the sum and difference frequencies of the two original waves. Or, mathematically put:

$$\sin A \times \sin B = \tfrac{1}{2}\{\cos(A - B) - \cos(A + B)\}$$

Inside the spectrum analyser, because the output of the multiplier stage is low-pass filtered, at all practical frequencies the sum frequencies disappear leaving only the difference frequency. And this will only be a steady DC signal ($A = B = 0\,\mathrm{Hz}$) when the two frequencies (the input signal or target signal and the search frequency or basis function) are exactly the same. Figure 4.11 illustrates this; only $(\sin x) \times (\sin x)$ results in a waveform which is asymmetrical about the x axis. In this manner, the single, component sine wave frequencies within a complex input waveform may be sought out by selecting different frequency basis functions and noting the DC voltage resulting from the multiplication voltage followed by low-pass filtering. (Although I have explained this in terms of voltages, clearly the same principal obtains in a digital system for symbolic numerical magnitudes.)

Phase

Actually the scheme described so far, though it often represents the complete system within a spectrum analyser, doesn't give us all the information we need to know about a waveform in the frequency domain. So far we only have a measure of the magnitudes of each of the sine wave frequencies present within the input signal. In order to reconstruct a waveform from the frequency domain description, we need to know the phases of each of the frequencies present within the original signal. It would be quite possible to do this by constructing some form of calibrated phase-shifting arrangement and adjusting for maximum 0 Hz output, once this had been found, using the technique above. Noting the phase value would yield all the information required for a complete frequency domain description. However, this isn't done. Instead, a technique is used whereby the input signal is multiplied by a sine basis function and a cosine basis function. These two functions can be thought of as separated by 90 degrees in phase. If you look at Figure 4.11 again, you'll notice that $(\sin x) \times (\sin x)$ produces a continuous offset whereas $(\sin x) \times (\cos x)$ does not. Whatever the phase of the input signal, it will generate a result from one or other (or both) of the multiplication processes. By knowing the magnitude of both the sine and cosine multiplications, it's possible to calculate the true magnitude of the original frequency component signal by calculating the square root of the sum of the squares of the two results; also its phase, because the tangent of the phase angle = result1/result2. With that intuitive explanation under our belts, let's look at the maths!

The frequency domain description of the signal completely specifies the signal in terms of the amplitude and phases of the various sinusoidal frequency components. Any signal $x(t)$, expressed as a function of time – in the so called *time domain* – can be instead expressed in the *frequency domain* $x(\omega)$, in terms of its frequency spectrum. The continuous-time Fourier integral provides the means for obtaining the frequency-domain representation of a signal from its time-domain representation and vice versa. They are often written like this:

Fourier transform:

$$X(\omega) = \int_{-\infty}^{+\infty} x(t)e^{-j\omega t}\,\mathrm{d}t$$

Inverse Fourier transform:

$$x(t) = \frac{1}{2\pi} \int_{-\infty}^{+\infty} X(\omega)e^{j\omega t}\,\mathrm{d}\omega$$

where $x(t)$ is a time-domain signal, $x(\omega)$ is the complex Fourier spectrum of the signal and (ω) is the frequency variable. This may look pretty scary, but take heart! The quoted exponential form of Fourier's integral is derived from Euler's Formula:

$$\exp jA = \cos A + j \sin A$$

The alternative form is found by replacing A by $-A$:

$$\exp(-jA) = \cos(-A) + j \sin(-A)$$

which is equivalent to

$$\exp(-jA) = \cos A - j \sin A$$

so it's really just a shorthand way of writing the process of multiplication by both sine and cosine basis functions and performing the integration which represents the equivalent of the subsequent low-pass filtering mentioned above.

An important difference between analogue and digital implementations is that in the former, the search (basis) function is usually in the form of a frequency sweep. The digital version, because it is a sampled system, only requires that the basis function operate at a number of discrete frequencies; it is therefore known as the *discrete Fourier transform*, or DFT. The fast Fourier transform is just a shortcut version of the full DFT. An incidental benefit of the discrete Fourier transform is that the maths is a bit easier to grasp. Because time is considered to be discrete (or non-continuous) in a digital system, the complication of conceptualizing integrals which go from the beginning to the end of time or from infinitely negative frequency to infinitely high frequency can be dispensed with! Instead the process becomes one of discrete summation: So, if we have a digital signal and we wish to discover the amplitudes and phases of the frequencies that comprise a time domain signal $x[n]$, we can calculate it thus:

$$a_k = \sum_{n=0}^{N-1} x(n) e^{(-j2\pi kn/N)}$$

where a_k represents the kth spectral component and N is the number of sample values in each period. Now, if you go to the trouble of calculating the DFT of a digital signal, the mathematics yields not only a zero-order spectrum, as we expect from the continuous transform, but a series of spectra centred on multiples of the sampling frequency. However, look again at Chapter 3, where we discovered that the spectrum of a sampled signal is exactly of this type. Despite the slightly counter-intuitive results of the DFT, you can see that its results are indeed valid in the sampled domain.

Windowing

If you imagine looking at a continuous 10 kHz sine wave on a spectrum analyser; you expect to see a single line at 10 kHz, a pure tone. But consider that you marvel at this sight for 1 minute before switching the signal generator off. In effect, the 10 kHz tone isn't pure at all; it has sidebands. Why? Because the 10 kHz carrier was modulated by a rectangular function of duration 1 minute. That is, it will have an infinite set of sidebands at spacings of 0.0167 Hz to the carrier. In fact, the only pure sine wave is one that started at the commencement of time and will continue to the end of time! Now this is mathematically satisfying, but it isn't very practical. What we need for everyday practical situations when we transform between the time and frequency domain is the ability to take a 'snapshot' through a mathematical 'window', which limits the transformation to a practical chunk of time or frequency. So persuasive is this metaphor that this manipulation is actually known as windowing.

The simplest form of windowing is a rectangular function in the time domain, like that illustrated in Figure 4.12, where the window is shown selecting a region of a sine wave waveform. If we transform the resulting signal (that is the sine wave signal multiplied by the rectangular window function), we can see what the resulting spectrum will look like. Now, the rectangular window function has a frequency spectrum of $\{(\sin x)/x\}$, and it is this spectrum that is evident in the transform, as shown in Figure 4.12. This result is obviously a bit confusing. What we need is a much gentler window function, which doesn't impress so much of itself on the transform. Ideally we need a rectangular function in the frequency domain, which (no prizes for guessing) translates to a $\{(\sin x)/x\}$ response in the time domain. Unfortunately this is very hard practically to achieve, so compromise window functions are used. These include the triangular window, the Hanning and Hamming window and the Blackman window, among others.

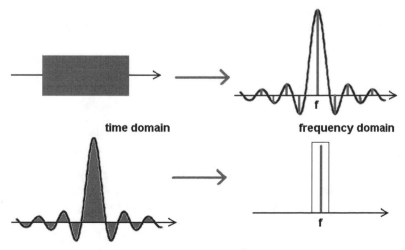

time domain   frequency domain

Figure 4.12 *Windowing of a continuous function*

2-D Fourier transforms

Two-dimensional (2-D) Fourier transforms were originally applied to image processing in applications as diverse as magnetic resonance imaging for medical applications and analysis of geophysical satellite images. The power of presenting image information in the spatial frequency domain lies in determining, for instance, a hidden periodicity that may not be discernible in the original image, and in the filtering techniques that may be applied to the transformed image before inverse transforming the image back into the spatial amplitude domain for viewing. The reasons why 2-D transforms are important in the context of digital television are different. In this case, transforms related to the Fourier transform (usually the discrete cosine transform) are used as a stage in data compression or 'source coding', as will be explained in Chapter 5.

In order to process multidimensional signals using the Fourier transform, the classical equations presented above need to be recast somewhat. Starting with the equation for the continuous transform above:

$$X(\omega) = \int_{-\infty}^{+\infty} x(t)\,e^{-j\omega t}\,dt \equiv X(f) = \int_{-\infty}^{+\infty} x(t)\,e^{-j2\pi ft}\,dt$$

we must first substitute the spatial variable x in place of t, and substitute the spatial frequency variable ν in place of f like this:

$$G(\nu) = \int_{-\infty}^{+\infty} g(x)\,e^{-j2\pi\nu x}\,dx$$

(Note that the functions are now G and g; this is arbitary and is only changed for clarification.)

Now we have an expression that allows us to analyse a signal that varies in space as opposed to a signal that varies in time.

However, an image varies spatially in two dimensions (pixels and lines, in television terms). Its continuous Fourier transform is obtained by evaluating a double integral:

$$G(\nu, v) = \iint_{-\infty}^{+\infty} g(x, y)\,e^{-j2\pi(\nu x + vy)}\,dx\,dy$$

which is pretty heavy-duty maths! Fortunately, as we saw above, discrete functions allow us a very considerable degree of simplification. The discrete form of the 2-D Fourier transform (2DFT) is:

$$G(\nu, v) \sum_{x=0}^{N-1}\sum_{y=0}^{N-1} g(x, y)\,e^{-j2\pi(\nu x + vy)/N}$$

where the image $g(x, y)$ is a function whose sampled and quantized amplitude is related to the brightness at a number (N) of discrete pixels (x, y) on the television screen, and $G(\nu, v)$ is a matrix of the spatial frequency transformation.

Mathematically, the double summation can be broken down into two stages; effectively two one-dimensional transforms performed one after the other. First an intermediate matrix $G(\nu, y)$ is derived by transforming each of the N rows like this:

$$G(\nu, y) = N \left\{ \frac{1}{N} \sum_{x=0}^{N-1} g(x, y) \, e^{-j2\pi\nu x/N} \right\}$$

Then the final matrix, $G(\nu, v)$, is computed by taking the DFT of each column of the intermediate matrix $G(\nu, y)$ like this:

$$G(\nu, v) = \frac{1}{N} \sum_{y=0}^{N-1} G(\nu, y) \, e^{-2\pi v y/N}$$

More about digital filtering and signal processing

The first computers, including those developed at Harvard University, had separate memory space for program and data, this topology being known as Harvard architecture. In fact the realization, by John von Neumann, the Hungarian born mathematician, that program instructions and data were only numbers and could share the same 'address-space', was a great breakthrough at the time and was sufficiently radical that this architecture is often named after its inventor. The advantage of the von Neumann approach was great simplification but at the expense of speed because the computer can only access either an instruction or data in any one processing clock cycle. The fact that virtually all computers follow this latter approach illustrates that this limitation is of little consequence in the world of general computing.

However, the speed limitation 'bottleneck' inevitable in the von Neumann machine, can prove to be a limitation in specialist computing applications like digital audio signal processing. As we have seen, in the case of digital filters, digital signal processing contains many, many multiply and add type instructions of the form

$$A = B.C + D$$

Unfortunately, a von Neumann machine is really pretty inefficient at this type of calculation so the Harvard architecture lives on in many DSP chips, meaning that a multiply and add operation can be performed in one clock cycle. This composite operation is termed a Multiply ACcumulate (MAC) function. A further distinction pertains to the incorporation within the

DSP IC of special registers which facilitate the managing of circular buffers.

The remaining differences between a DSP device and a general purpose digital microcomputer chip relate to the provision of convenient interfaces, thereby allowing direct connection of ADCs, DACs and digital transmitter and receiver ICs.

Convolution

In the simple three-stage digital filter considered above, we imagined the step function being multiplied by a quarter, then by a half and finally by a quarter again: at each stage, the result was added up to give the final output. As we saw with digital image processing, this actually rather simple process is given a frightening name in digital signal processing theory, where it is called convolution. You'll remember that discrete convolution is a process which provides a single output sequence from two input sequences. In the example above, a time-domain sequence – the step function – was convolved with the filter response yielding a filtered output sequence. In textbooks convolution is often denoted by the character '$*$'. So if we call the input sequence $h(k)$ and the input sequence $x(k)$, the filtered output would be defined as

$$y(n) = h(k) * x(k)$$

Impulse response

A very special result is obtained if a unique input sequence is convolved with the filter coefficients. This special result is known as the filter's impulse response and the derivation and design of different impulse responses is central to digital filter theory. The special input sequence we use to discover a filter's impulse response is known as the 'impulse input'. (The filter's impulse response being its response to this impulse input.) This input sequence is defined to be always zero, except for one single sample which takes the value 1 (i.e. the full-scale value). We might define, for practical purposes, a series of samples like this:

$$0, 0, 0, 0, 0, 1, 0, 0, 0, 0$$

Now imagine these samples being latched through the three-stage digital filter shown above. The output sequence will be:

$$0, 0, 0, 0, 0, 1/4, 1/2, 1/4, \ 0, 0, 0, 0$$

Obviously the zeros don't really matter: what's important is the central section: 1/4, 1/2, 1/4. This pattern is the filter's impulse response.

FIR and IIR filters

Notice that the impulse response of the filter above is finite: in fact, it only has three terms. So important is the impulse response in filter theory that this type of filter is actually defined by this characteristic of its behaviour and is named a finite impulse response (FIR) filter. Importantly, note that the impulse response of an FIR filter is identical to its coefficients.

Now look at the digital filter in Figure 4.13. This derives its result from both the incoming sequence and from a sequence which is fed back from the output. Now if we perform a similar thought experiment to the convolution example above and imagine the resulting impulse response from a filter of this type, it results in an output sequence like that illustrated in the figure: that's to say, an infinitely decaying series of values. Once again, so primordial is this characteristic, that this category of filter is termed an infinite impulse response (IIR) filter.

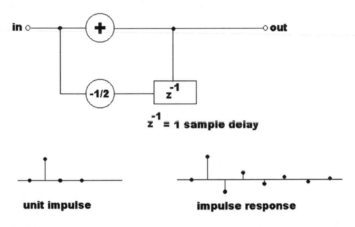

Figure 4.13

IIR filters have both disadvantages and advantages over the FIR type. First, they are very much more complicated to design; because their impulse response is not simply reflected by the tap coefficients, as in the FIR. Secondly, it is in the nature of any feedback system (like an analogue amplifier), that some conditions may cause the filter to become unstable if it is has not been thoroughly designed, simulated and tested. Furthermore, the inherent infinite response may cause distortion and/or rounding problems as calculations on smaller and smaller values of data are performed. Indeed, it's possible to draw a parallel between IIR filters and analogue filter circuits: they share the disadvantages of complexity of design and possible instability and distortion, but they also share the great benefit that they are efficient. An IIR configuration can be made to implement complex filter functions with only a few stages, whereas the equivalent FIR filter would require many hundreds of taps with all the drawbacks of cost and signal delay that this implies. (Sometimes FIR and IIR filters are referred to as 'recursive' and 'non-recursive' respectively: these terms directly reflect the filter architecture.)

Design of digital filters

Digital filters are nearly always designed from a knowledge of the required impulse response. IIR and FIR filters are both designed in this way; although the design of IIR filters is complicated because the coefficients do not represent the impulse response directly. Instead, IIR design involves various mathematical methods which are used to analyse and derive the appropriate impulse response from the limited number of taps. This makes the design of IIR filters from first principles rather complicated and maths-heavy! (Nonetheless, I'll show how simple filters may be easily designed later in the chapter.) FIRs are easier to understand and are a good place to start because even a brief description gives a good deal of insight into the design principles of all digital filters.

We already noted that the response type of the 1/4, 1/2, 1/4 filter was a low-pass; remember it 'slowed-down' the fast rising edge of the step waveform. If we look at the general form of this impulse response, we will see that this is a very rough approximation to the behaviour of an ideal low-pass filter which we already met in relation to reconstruction filters. There we saw that the $(\sin x)/x$ function defines the behaviour of an ideal, low-pass filter and the derivation of this function is given in Figure 4.14. Sometimes termed a sinc function, it has the characteristic that it is infinite, gradually decaying with ever smaller oscillations about zero. This illustrates that the perfect low-pass FIR filter would require an infinite response, an infinite number of taps and the signal would take an infinitely long time to pass through it! Fortunately for us, we do not need such perfection.

However, the 1/4, 1/2, 1/4 filter is really a very rough approximation indeed. So let's now imagine a better estimate to the true sinc function and design a relatively simple filter using a 7-tap FIR circuit. I have derived the values for this filter in Figure 4.14. This suggests a circuit with the following tap values:

$$0.3, \ 0.6, \ 0.9, \ 1, \ 0.9, \ 0.6, \ 0.3$$

The only problem these values present is that they total to a value greater than 1. If the input was the unity step-function input, the output would take on a final amplitude of 4.6. This might overload the digital system, so we normalize the values so that the filters response at DC (zero frequency) is unity. This leads to the following, scaled values:

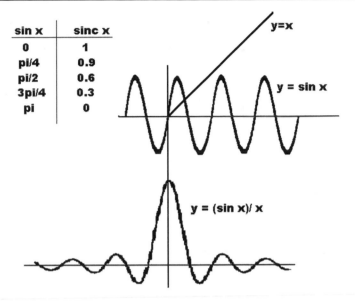

sin x	sinc x
0	1
pi/4	0.9
pi/2	0.6
3pi/4	0.3
pi	0

Figure 4.14   *Derivation of sin x/x response*

$$0.07, 0.12, 0.2, 0.22, 0.2, 0.12, 0.07$$

The time-domain response of such an FIR filter to a step and impulse response is illustrated in Figure 4.15. The improvement over the three-tap filter is already obvious.

Frequency response

But how does this response relate to the frequency domain response? For it is usually with a desired frequency response requirement that filter design begins. To design a 'windowed-sinc' filter (for that is what a filter with a limited sin x/x response is technically called), two parameters must be selected:

f_c = the 'cut-off' frequency
M = length of the kernal.

(f_c) is expressed as a fraction of sampling rate, and must therefore be between 0 and 0.5.

M sets the rate of 'cut-off' according to the approximation:

$$M \approx \frac{4}{T}$$

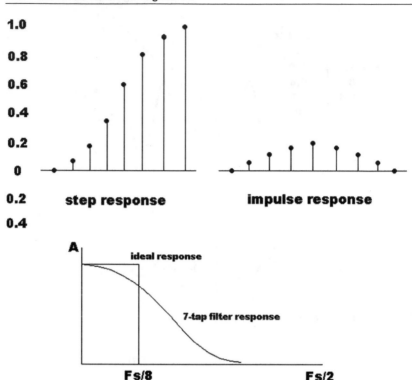

Figure 4.15   *Response of simple, digital, low-pass filter (see text)*

Where T is the width of the 'transition band'. This transition band is also expressed as a fraction of the sampling frequency (and must therefore be between 0 and 0.5).

The transition band (T) is measured between amplitude response points for 99 per cent and 1 per cent. Note this is not the normal definition from analogue practice and neither is the cut-off frequency. In analogue practice this is defined as the point in which the amplitude response falls to −3 dB. In digital practice, it is defined as the point in which the amplitude response has fallen to −6 dB.

Having decided on f_c and T, and thereby M, the individual terms of the filter are calculated according to the equation

$$b[i] = K \frac{\sin 2\pi fc(i - \frac{M}{2})}{i - \frac{M}{2}} \left[0.42 - 0.5 \cos\left(\frac{2\pi i}{M}\right) + 0.08 \cos\left(\frac{4\pi i}{M}\right) \right]$$

This may look a bit ghastly but it is actually quite straightforward when 'plugged into' a programmable calculator or a computer.

There are one or two things to watch out for:

1. *K*, which is a normalization constant, should be chosen so that all the terms add to unity. This assumes no gain at DC. (Actually, how can you know this before you start? You can't! So the best thing is to ignore this term and calculate it at the end of the process, once all the kernal coefficients have been calculated, so as to give an overall gain of one at DC.)
2. *M* must be chosen as an even number.
3. The sample number *i* runs from zero to *M*/2, so the filter will eventually have an odd number of terms, *M* + 1.
4. The centre of f_c sinc function will provoke your computer to 'melt-down' because the term $(i - M/2)$ will be zero; resulting in a 'division by zero' error. To avoid this, the term $i = M/2$ should be calculated thus:

$$b[i] = 2\pi fcK, \text{ when } i = \frac{M}{2}$$

Derivation of band-pass and high-pass filters

All digital filters start life as low-pass filters and are then transformed into band-pass or high-pass types. A high-pass is derived by multiplying each term in the filter by alternating values of +1 and −1. So, our simple low-pass filter above,

$$0.07, 0.12, 0.2, 0.22, 0.2, 0.12, 0.07$$

is transformed into a high-pass like this:

$$+0.07, -0.12, +0.2, -0.22, +0.2, -0.12, +0.07$$

The impulse response and the frequency of this filter is illustrated in Figure 4.16. If you add up these high-pass filter terms, you'll notice they come nearly to zero. This demonstrates that the high-pass filter has practically no overall gain at DC, as you'd expect. Notice too how the impulse response looks 'right'; in other words, as you'd anticipate from an analogue type.

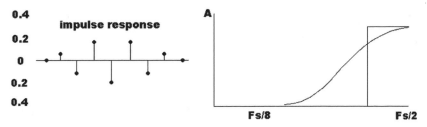

Figure 4.16 *Response of a simple, derived, high-pass filter (see text)*

A band-pass filter is derived by multiplying the low-pass coefficients with samples of a sine wave at the centre frequency of the band-pass. Let's take our band-pass to be centred on the frequency of $F_s/4$. Samples of a sine wave at this frequency will be at the 0 degree point, the 90 degree point, the 180 degree point and the 270 degree point and so on. In other words,

$$0, 1, 0, -1, 0, 1, 0, -1$$

If we multiply the low-pass coefficients by this sequence we get the following:

$$0, 0.12, 0, -0.22, 0, +0.12, 0$$

The impulse response of this circuit is illustrated in Figure 4.17. This looks intuitively right too, because the output can be seen to 'ring' at $F_s/4$, which is what you'd expect from a resonant filter. The derived frequency response is also shown in the diagram.

Designing an IIR filter

Infinite impulse response (IIR) filters are practically very simple, but the maths can be very complicated indeed!

The essence of an IIR filter or 'recursive' filter is the 'recursion equation':

$$y[n] = a_0 x[n] + a_1 x[n-1] + a_2 x[n-2] + b_1 y[n-1] + b_2 y[n-2]$$

where output is $y[n]$, derived from the input sequence $x[n]$ and the output sequence. It is this feature which makes the IIR so powerful because, in effect, IIR filters convolve the input signal with a very long filter kernal; although only a few coefficients are involved.

The relationship between coefficients and the filter's response is given by a mathematical technique known as the z-transform. Happily, some IIRs can be designed and understood without an understanding of the z-transform, which is very mathematical and rather complicated (see below).

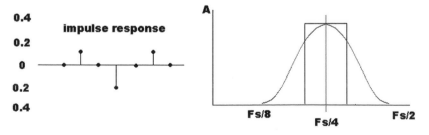

Figure 4.17   *Response of a simple, derived, band-pass filter*

The design of IIR filters is often based on analogue practice. The simplest place to start is therefore the simple low-pass RC (simple-pole) filter. This is modelled in the IIR filter by the relation

$$y[n] = (1 - \alpha)x[n] + \alpha \cdot y[n-1]$$

Where α controls the characteristics of the filter (and α can take any value between zero and 1).

The value of α can be computed from the equivalent time constant of the analogue filter (RC in seconds). This time constant is defined, in a digital filter, by

$$d,$$

where d = number of samples it takes a digital filter to reach 63 per cent of its final value when presented with a step function at its input, such that

$$\alpha = e^{\frac{-1}{d}} \quad \text{or} \quad d = -\left(\frac{1}{\ln \alpha}\right)$$

The cut-off frequency (f_c) is defined by the expression

$$\alpha = e^{-2\pi f_c}$$

where f_c is a valve between 0 and 0.5 (i.e. a fraction of sampling frequency).

The high-pass equivalent of the simple RC low-pass filter is modelled like this:

$$y[n] = (1 + \alpha)x[n]/2 - (1 + \alpha)x[n-1]/2 + \alpha \cdot y[n-1]$$

α is defined in exactly the same way.

IIR filter design example

We are asked to duplicate, in the digital domain, the analogue, single-pole, low-pass filter formed by a series 8.2k and a shunt 10 nF to ground. We can calculate the f_c for this as

$$\frac{1}{2 \cdot \pi \cdot 8.2 \times 10^3 \cdot 10 \times 10^{-9}} = 1940 \text{ Hz}$$

$$RC = (8.2 \times 10^3) \times (10 \times 10^{-9}) = 8.2 \times 10^{-5} = 82 \text{ } \mu S$$

In the digital domain, our sampling frequency is to be professional 48 kHz, meaning that the audio is sampled at

$$\frac{1}{48\,000} = 20.8 \text{ } \mu S$$

We can therefore calculate the all-important factor α by way of d:

$$d = \frac{82 \ \mu S}{20.8 \ \mu S} = 3.94$$

$$\alpha = e^{\frac{-1}{3.94}}$$

$$\alpha = 0.77$$

Alternatively, we can use f_c to calculate α:

f_c (as a fraction of sampling frequency),

$$\frac{1940}{48\,000} = 0.04$$

$$\alpha = e^{-2\pi(0.04)}$$

$$\alpha = 0.77$$

This gives us the recursion example:

$$y[n] = 0.23 \times [n] + 0.77y[n-1]$$

Table 4.1 shows the calculation of the first few samples resulting from this recursive filter when a step function is applied (x is input, y is output).

Note the fourth term following the step input on the input (marked with *): this confirms our calculations that it is approximately at the fourth sample that the output will reach the definitive 63 per cent point.

The digital filter could be implemented in software or hardware: a hardware implementation is illustrated in Figure 4.18.

Table 4.1

X	Y
0	0
0	0
1	0.23
1	0.41
1	0.54
1	0.65*
1	0.73
1	0.79
1	0.84
1	0.88
1	0.9

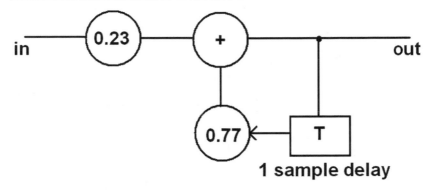

Figure 4.18 *Practical single-pole, low-pass IIR filter*

High-pass filter example

There is often a need for a high-pass filter with an f_c at a very low frequency, for removing DC offset. In this example we will use a simple, single-pole, high-pass, IIR filter to perform this task.

Once again we will start with the choice of f_c which we will use to calculate α. But this time we will choose a frequency which will give us a very smooth function of the 48 kHz sampling frequency: 12 Hz. This is low enough to preserve the audio signal.

$$\alpha = e^{-2\pi f_c}$$

where $f_c = 12/48\,000 = 0.00025$, $\alpha = 0.998$ and $d = 500$ samples.

Substituting these figures into the earlier equation for the high-pass filter,

$$y[n] = (1 + 0.998)x[n]/2 - (1 + 0.998)x[n-1]/2 + 0.998y[n-1]$$

or

$$y[n] = 0.999x[n] - 0.999x[n-1] + 0.998y[n-1]$$

Now imagine the step function as we did before. From inspection, it is clear that the first term $y[n]$ will equal $0.999 \times [n]$, but all subsequent terms will rely only on the feedback term (because the input remains the same and the first two terms will simply cancel out). The resulting decay is governed by α in the feedback term. It is therefore important to note that:

$$(0.998)^{500} = 0.368 \text{ or approximately 37 per cent}$$

which means that the filter response will have decayed by 63 per cent in 500 samples; the definition of the time constant RC.

In hardware, the filter would look like Figure 4.19.

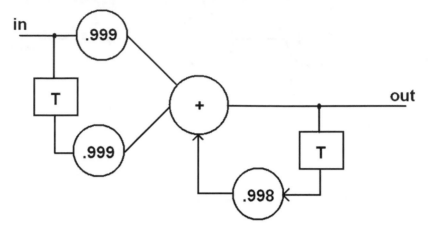

Figure 4.19   *Practical single-pole, high-pass IIR filter*

Digital frequency domain analysis – the z-transform

The z-transform of a digital signal is identical to the Fourier transform except for a change in the lower summation limit. In fact, you can think of 'z' as a frequency variable which can take on real and imaginary (i.e. complex) values. When the z-transform is used to describe a digital signal, or a digital process (like a digital filter) the result is always a rational function of the frequency variable z. That's to say, the z-transform can always be written in the form

$$X(z) = N(z)/D(z) = K(z - zl)/(z - pl)$$

Where the z's are known as 'zeros' and the p's are known as 'poles'.

A very useful representation of the z-transform is obtained by plotting these poles and zeros on an Argand diagram, the resulting two-space representation being termed the 'z-plane'. When the poles and zeros are plotted in this way, they give us a very quick way of visualizing the characteristics of a signal or digital signal process.

Problems with digital signal processing

As we have already seen, sampled systems exhibit aliasing effects if frequencies above the Nyquist limit are included within the input signal. This effect is usually no problem because the input signal can be filtered so as to remove any offending frequencies before sampling takes place. However, consider the situation in which a band-limited signal is subjected to a non-linear process once in the digital domain. This entirely digital processes generates a new large range of harmonic frequencies (just like its analogue counterpart), as shown in Figure 4.20. The problem arises that many of these new harmonic frequencies are actually above the half-sampling frequency limit and get folded back into the pass-band as illustrated in Figure 4.21.

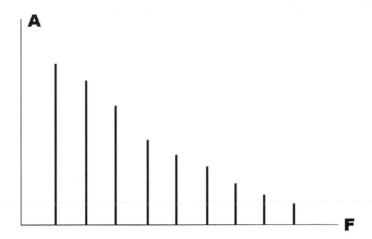

Figure 4.20   *A signal with its harmonics in the analogue domain*

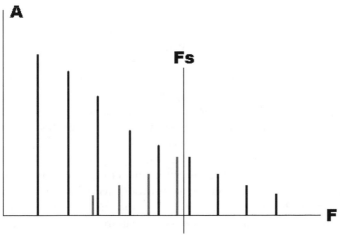

Figure 4.21   *The harmonics may be 'folded back' in the digital domain*

5
Video data compression

Basic concepts

Digital television in the home was considered an impossibility until very recently. That it is possible now is entirely due to the incredible advances in the suite of data-compression techniques we call MPEG-II. This chapter describes the compression tools used in contemporary picture coding.

Entropy, redundancy and artefacts

If I say to you, 'Wow, I had a bad night, the baby cried from three 'til six!' you understand perfectly what I mean because you know what a baby crying sounds like. I might alternatively have said, 'Wow, I had a bad night, the baby did this; wah, bwah, bwah, wah ...' and continue for 3 hours. Try it – you'll find a lot of friends (!) because nobody needs to have it demonstrated. Most of the 3-hour impersonation is superfluous. The second message is said to have a high level of redundancy in the terms of communication theory. The trick performed by any compression system is sorting out the necessary information content – sometimes called the entropy – from the redundancy. (If, like me, you find it difficult to comprehend the use of the term 'entropy' in this context, consider this: Entropy refers here to a lack of pattern; to disorder. Everything in a signal that has a pattern is, by definition, predictable and therefore redundant. Only those parts of the signal that possess no pattern are unpredictable and therefore represent necessary information.)

All compression techniques may be divided into lossless systems and lossy systems. Lossless compression makes use of efficiency gains in the manner in which the data is coded. All that is required to recover the original data exactly is a decoder, which implements the reverse process performed by the coder. Such a system does not confuse entropy for redundancy and hence dispense with important information. However, neither does the lossless coder perfectly divide entropy from redundancy. A good deal of redundancy remains, and a lossless system is therefore only capable of relatively small compression gains. Lossy compression techniques attempt a more complete distinction between entropy and redundancy by relying on a knowledge of the predictive powers of the human perceptual systems. This explains why these systems are referred to as implementing perceptual coding techniques. Unfortunately, not only are these systems inherently more complicated, they are also more likely to get things wrong and produce artefacts.

Lossless compression

Consider the following contiguous stream of luminance bytes taken from a bitmap graphic:

00101011
00101011
00101011
00101011
00101011
00101011
00101100
00101100
00101100
00101100
00101100

There must be a more efficient way of coding this! 'Six lots of 00101011 followed by five lots of 00101100' springs to mind. Like this:

00000110
00101011
00000101
00101100

This is the essence of a compression technique known as run-length encoding (RLE). RLE works really well, but it has a problem; if a data file is comprised of data that is predominantly non-repetitive data, RLE actually makes the file bigger! RLE must therefore be made adaptive so that it is only applied to strings of similar data (where redundancy is high), and when the coder detects continuously changing data (where entropy is high) it simply reverts back to sending the bytes in an uncompressed form. Evidently it also has to insert a small information overhead to instruct the decoder when it is (and isn't) applying the compression algorithm.

Another lossless compression technique is known as Huffman coding, and is suitable for use with signals in which sample values appear with a known statistical frequency. The analogy with Morse code is often drawn, where letters that appear frequently are allocated simple patterns and letters that appear rarely are allocated more complex patterns. A similar technique, known by the splendid name of the Lempel-Ziv-Welch (LZW) algorithm, is based on the coding of repeated data chains or patterns. A bit like Huffman's coding, LZW sets up a table of common patterns and code-specific instances of patterns in terms of 'pointers' that refer to much longer sequences in the table. The algorithm doesn't use a pre-defined set of patterns, but instead builds up a table of patterns which it 'sees' from the incoming data. (The GIF image-file format employs this type of compression.)

LZW is an effective technique – even better than RLE. But for the really high compression ratios made necessary by the transmission and storage of television pictures down low bandwidth links, different approaches are required. These are described below.

De-correlation

For real audio and video signals, Huffman (or even LZW) coding is less effective than we might hope because the statistical use of each coding level is roughly equal. The situation is therefore unlike Morse code, where it is clear that some letters are used more frequently than others. We consequently rely on a technique that may be used, termed de-correlation, in which the original source material is subjected to a known de-correlation function before Huffman coding, the idea being to concentrate the image over fewer codes, thereby making subsequent Huffman coding more worthwhile. When reconstructing the waveform, the incoming data must – of course – be passed through a compensating correlation function.

The simplest example of this technique is the use of a straightforward one-sample delay and subtractor, as illustrated in Figure 5.1. Imagine the image shown in Figure 5.2 subjected to such a filter. The resulting image is shown in Figure 5.3. As you would expect, large plain surfaces (where adjacent pixels are similar) produce an output that is frequently at very low level or even zero. By contrast, edges and corners, where there is a large difference between adjacent pixels, produce large values. These, however, are relatively infrequent. This transformation thereby yields a signal with widely differing statistics; frequent samples at low level and infrequent samples at high levels, which is ideal for Huffman type coding. In fact, Figure 5.1 demonstrates a widely used coding technique known as DPCM (differential pulse code modulation).

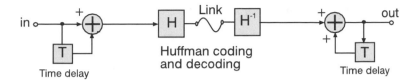

Figure 5.1 *Differential pulse code modulation (DPCM)*

Figure 5.2 *A photographic image*

Figure 5.3 *The image in Figure 5.2 subjected to DPCM coding process*

Lossless DPCM and lossy DPCM

If we redraw Figure 5.1 as shown in Figure 5.4, with the delay replaced with a block entitled 'predictor', we can see an important principle at work. Imagine that the predictor was perfect; that it 'knew' exactly what the next sample would be. Clearly it wouldn't be necessary to send any information by means of the link at all, because the predictor at the receiving end would be able to create the new sample from what it knew about the previous samples! Similarly, the predictor at the sending end would have predicted exactly the right value and thus the result of the subtraction process would be zero.

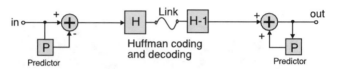

Figure 5.4   *Predictive DPCM*

Of course, a perfect predictor isn't possible! But the better the predictions are, the smaller is the difference between the input signal and the prediction; and thus the lower the eventual data rate over the link. In fact, the data over the link (since it is that which differs from a perfectly predicted signal) is referred to as prediction error.

In an attempt to reduce overall data-rate, we can add a quantizer to the lossless DPCM system as illustrated in Figure 5.5, thereby literally limiting the possible gamut of prediction error signals. Unfortunately this doesn't work at all well, because the two 'ends' of the system are now no longer the same and prediction error-signal quantization errors can build up in a catastrophic way. This is solved by effectively deriving the error signal from the receiver's predictor (rather than the transmitter's predictor). A real receiver is not used; instead a facsimile receiver is incorporated in the transmitter as shown in Figure 5.6. Resulting errors are thereby limited to those arising in the link. This system is known as error-containing lossy DPCM.

When a system must be designed to cope with link errors, the predictor is often deliberately arranged to be non-optimal because, although this

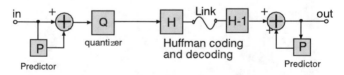

Figure 5.5   *Quantized predictive DPCM – results in build-up of prediction error*

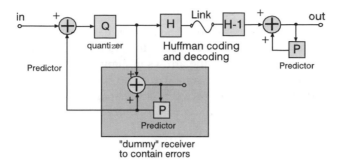

Figure 5.6   *Error-contained DPCM*

obviously creates a higher overall data-rate, it ensures there is always some transmitted data to correct for erroneous data on the loop – thus limiting the lifetime of the resulting decoding error. The non-optimal predictor is sometimes referred to as a 'leaky' predictor.

Frame differences and motion compensation

Whilst a single sample (pixel) delay can work quite well for image compression, even better results are obtained if the difference signal is derived between successive frames of video. This is the essence of most powerful video compression schemes (MPEG included); that differences in frames are sent rather than information 'frame by frame'.

Difference information, of course, works really well for still pictures, but the necessary data rate climbs when there is movement in the frame, as shown in Figure 5.7. This is a shame because, as Figure 5.7 shows, most of

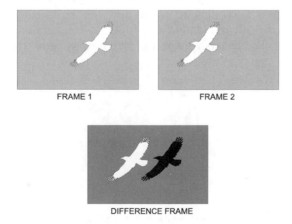

Figure 5.7   *Motion in successive frames of video*

the information in Frame 2 exists in Frame 1; it's just that some information has been geometrically translated. The answer to this is to send information about movement by means of vectors notating the movement of blocks of pixels, illustrated by the arrows in Figure 5.8. If this information is wholly accurate (as it is assumed for the idealized situation in the figure), then the difference between frames again becomes zero, as illustrated. This technique is known as *motion compensation.*

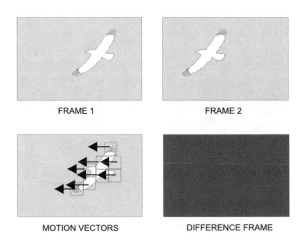

FRAME 1   FRAME 2

MOTION VECTORS   DIFFERENCE FRAME

Figure 5.8   *Motion coded as vectors indicating the movement of blocks of pixels*

But how is movement detected? There are a number of techniques, but the simplest and most widely used is block matching. The principle is illustrated in Figure 5.9, in which a block of picture is compared with a number of possible positions within the subsequent frame; the resulting match is ranked according to score and the best match taken as a candidate vector. All movement detection is computationally intensive, and there are usually incorporated in practical systems some means of limiting the searches to areas containing movement.

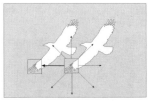

BLOCK SEARCH

Figure 5.9   *Mechanism of block-matching used to derive vector information*

You might think that motion compensation is a dangerous business; suppose, for example, that a block is incorrectly assigned and the bird's beak is suddenly assigned – incorrectly – to a wing-tip position! Surely a visual anomaly would result which would be laughable? In fact, motion compensation for compression is not as critical as it is for other applications because (once the vectors have been calculated) the difference information between frames is calculated again and is sent as normal. In other words, motion compensation can be thought of solely as a system to reduce differential coded bit-rate. If there is a catastrophic failure of motion compensation, it simply results in a momentary greater bit-rate in the frame-difference amplitude information.

Fourier transform-based methods of compression

Look at the series of pictures in Figure 5.10. In an attempt to try to reduce the amount of data, I've simply truncated the number of luminance bits. The first picture has 8-bit luminance values, the second has 7-bit, the third, 5-bit. The fourth and fifth have 3-bit and 2-bit values respectively. Notice that, even at 6 bits, the image has taken on contours like a survey map. It looks very unnatural, but see how the details of the ironwork in the bridge remain distinguishable even in the crudest image. This technique of reducing the resolution of individual samples is obviously useless, but it reveals an important phenomenon; the gradual shadings, which are the hallmark of photographic images, are characterized by adjacent pixels that vary by very small steps. Consequently, any coarsening of quantization levels distorts these gradual changes in the manner illustrated. These gradual changes, spread over many pixels, are the low spatial-frequency components within the image (low-frequency waves vary their amplitude slowly). Details of the ironwork, however, which vary markedly from pixel to pixel, are the high spatial-frequency components of the image and, as we can see, these are not as sensitive to reductions in resolution. What is required is a method that permits us to code low frequencies with more resolution than high frequencies. This approach is the one taken by both the JPEG photographic image compression system and as part of the MPEG system, which was developed for moving images.

Transform coding

How is the separation of low and high frequencies accomplished? The solution, in most current forms of image data compression, is the discrete cosine transform – a mathematical 'cousin' of the 2-D Fourier transform described in the previous chapter. This is described below. In order to see how transform coding is achieved, a much simpler transform is introduced first.

(a)

(b)

Figure 5.10   (a) *Photographic image – 8 bit representation;* (b) *Photographic image – 6 bit representation;* (c) *Photographic image – 5 bit representation;* (d) *Photographic image – 3 bit representation;* (e) *Photographic image – 2 bit representation*

Take the image in Figure 5.2. Let's imagine that we transform two pixels at a time using what I shall call the Brice transform. The Brice transform (for which I shall not be receiving accolades as a mathematician!) is defined thus: each adjacent pair of pixels (P1, P2) creates a transform pair of the form:

$$T1 = \{0.5(P1 + P2)\}$$

and

$$T2 = (P1 - P2)$$

In words, one of the transform pair is an average of two pixels and the other is the difference between them. You can't get much simpler than that! The process is illustrated in Figure 5.11a.

Also in Figure 5.11a the inverse Brice transform is performed, and you will see that the mathematical manipulation is most conveniently ex-

(c)

(d)

(e)

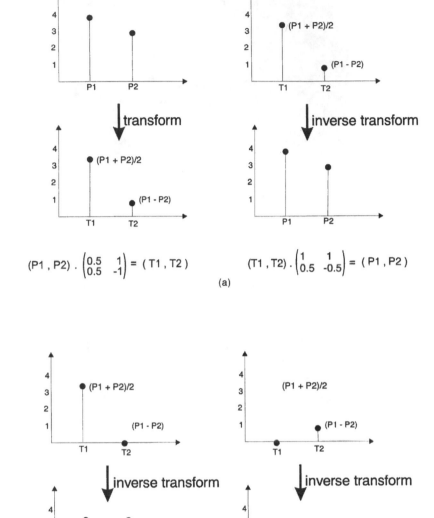

Figure 5.11   (a) *The Brice transform;* (b) *The effect of setting each of the transformed terms to zero*

pressed in terms of matrix multiplication. (See Chapter 8 for a revision of matrix maths.)

The Brice transform:

$$(T1, T2) = (P1, P2) \cdot \begin{pmatrix} 0.5 & 1 \\ 0.5 & -1 \end{pmatrix}$$

The inverse Brice transform:

$$(P1, P2) = (T1, T2) \cdot \begin{pmatrix} 1 & 1 \\ 0.5 & -0.5 \end{pmatrix}$$

I hope you can see intuitively that an average will tend to prejudice against high frequencies and a difference function will prejudice against low frequencies. Proof of this is given in Figure 5.11b, in which T1 and T2 are each set to zero before the inverse Brice transform is performed: When T2 is set to zero, it is the constant luminance value that is preserved (the DC term); in other words, all the AC content has been eradicated. When T1 is set to zero, the DC term is eradicated, leaving just the AC part. Further proof is given in the images in Figures 5.12 and 5.13, which show each half of the transformation and its visual effect on the picture. Notice that the averaging part has reduced high frequencies (Figure 5.12), whereas the difference part has exaggerated high frequencies (Figure 5.13). In practice, in sampled digital systems all transforms are performed this way; by means of matrix multiplication over a group of samples, thereby yielding a group of transform coefficients. Similarly, all inverse transforms are performed the same way; with matrix multiplication of a group of coefficients to reproduce the original group of samples.

A practical mix

Gestalt is a branch of psychology that is popularized by the term that the power of the human mind is 'more than the sum of its parts'. The same philosophy applies aptly to image data compression. All powerful compression schemes rely of an eclectic mix of the above techniques to achieve the best results. Each technique is chosen to complement the others. Looking again at Figures 5.12 and 5.13, notice how the statistical distribution of pixel amplitudes has been changed by the averaging and difference-taking operations (each half of the Brice transform). In the original picture (Figure 5.14), there is a wide spread of values. In the semi-transformed pictures, the spread is much less. This shows that not only does the frequency domain transform aid matters in discriminating between high frequencies and low frequencies, thereby allowing us to quantize more coarsely the high frequencies; it also produces data that is much more amenable to further reduction by downstream lossless compression techniques like Huffman coding.

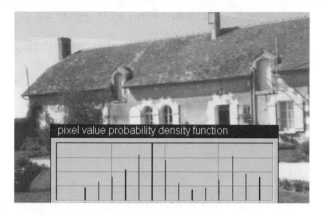

Figure 5.12 *First part of the Brice transform as applied to a real image*

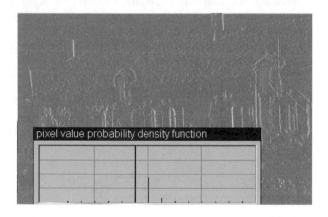

Figure 5.13 *Second part of Brice transform*

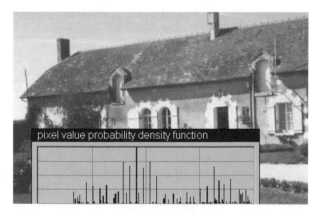

Figure 5.14 *Note the probability density function of original image pixel valued when compared with Figures 5.12 and 5.13*

JPEG

JPEG compression was developed by the Joint Photographic Experts Group of the International Standards Organization. Really it should be referred to as ISO10918, but the eponymous acronym has stuck. JPEG is a multiple pass compression technique, which has four distinct phases:

1 Colour–space conversion
2 Discrete cosine transform
3 Quantization
4 RLE and Huffman coding.

Colour–space conversion applies exactly the same technique as in television and is described in Chapter 2. To recap, remember that the eye is relatively insensitive to coloured detail. This is obviously a phenomenon that is of great relevance to any application of colour picture compression. So JPEG's first stab at data reduction is the axis transformation usually referred to as RGB to YUV conversion, and it is achieved by mathematical manipulation of the form:

$$Y = 0.3R + 0.59G + 0.11B$$

$$U = m(B - Y)$$

$$V = n(R - Y)$$

This transformation alone doesn't achieve any compression, but it was noted that, in television applications, the U and V signals may be a much lower bandwidth than the luminance signal without significant perceptible loss of quality. JPEG compression's answer to this is to down-sample (the same technique as used in the digital television CCIR 601 standard). This is the first lossy step in JPEG compression, and it consists of coding only one U and V colour difference value for every two (block of 2 × 1), four (block of 2 × 2) or even 16 (block of 4 × 4) luminance pixels, depending on the level of compression.

The next step in JPEG compression is to transform the image data to the frequency domain. The mathematical transform used to implement this is known as the discrete cosine transform (DCT). The DCT takes advantage of a distinguishing feature of the cosine function that is illustrated in Figure 5.15. Notice that the cosine curve is symmetrical about the time origin. In fact, it's true to say that any waveform which is symmetrical about an arbitrary 'origin' is made up of solely of cosine functions. Difficult to believe, but consider adding other cosine functions to the curve illustrated in Figure 5.15. It doesn't matter what size or what period waves you add, the curve will always be symmetrical about the origin. Now it would obviously be a great help when we come to perform a Fourier transform if we know the function to be transformed is only made up of cosines,

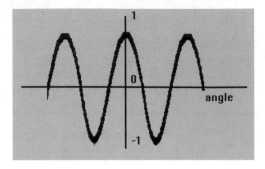

Figure 5.15   *The cosine function – note that it is symmetrical about zero*

because that would cut down the maths by half. This is exactly what is done in the DCT. A sequence of samples from the incoming waveform are stored and reflected about an origin, then one half of the Fourier transform performed. When the waveform is inverse transformed, the front half of the waveform is simple ignored, revealing the original structure. (Notice the DCT is sometimes referred to as a version of DFT in which all the terms in the frequency-domain are 'real' – meaning cosines.)

In the case of the JPEG compression system, the DCT is not performed on the whole image at a time. Although this is theoretically possible, the number and complexity computations would be enormous. Instead, the DCT is performed on blocks of 8 × 8 pixels. The results are stored in a similar 8 × 8 array – one for luminance and one each for colour difference signals. This is illustrated in Figures 5.16 and 5.17, where the values of a block of 8 × 8 pixels (Figure 5.16) are shown transformed to their representation after DCT (Figure 5.17).

The following step, quantization, is where the real compression happens, because the higher frequency components are quantized

255	0	255	0	255	0	255	0
0	255	0	255	0	255	0	255
255	0	255	0	255	0	255	0
0	255	0	255	0	255	0	255
255	0	255	0	255	0	255	0
0	255	0	255	0	255	0	255
255	0	255	0	255	0	255	0
0	255	0	255	0	255	0	255

Figure 5.16   *A chequered 8 × 8 block of black and white pixels*

32640	0	0	0	0	0	0	0
0	530.2	0	625.4	0	936	0	2665
0	0	0	0	0	0	0	0
0	625.4	0	737.7	0	1104	0	3144
0	0	0	0	0	0	0	0
0	936	0	1104	0	1652	0	4705
0	0	0	0	0	0	0	0
0	2665	0	3144	0	4705	0	13399

Figure 5.17   *The chequered block after DCT*

using fewer possible levels than the low frequencies. It is by controlling the scaling of the quantization levels at this stage that the different compression ratios are achievable within the JPEG algorithm. The DCT coefficients are serialized by scanning the coefficient array in a zig-zag pattern starting at the top left-hand corner and ending up at the right lower corner. Because so many coefficients are equal (and many zero!) after a typical quantized DCT, Huffman coding compresses the data still further. The table for the Huffman coding can either be a standard one or an image-specific one, in which case it must be sent with the file. The specific file approach results in a slightly higher compression ratio.

Motion JPEG (MJPEG)

One solution to coding television images is simply to apply JPEG coding to individual fields or frames. This is sometimes done, and the final result is termed motion JPEG or MJPEG coding. The disadvantage of JPEG-type image coding for television is that, for acceptable television pictures, it is still necessary to use somewhere in the region of 2 bits to code each pixel. This provides insufficient compression for television pictures for digital television broadcast applications. However, MJPEG and its cousins, such as DVC, DV-Cam and DVCPro (all frame-bound DCT based schemes), have a number of advantages which will probably mean that it will continue to play a role in studios and editing situations.

MPEG

The MPEG video compression used in digital television pursues the mix of compression techniques to its logical conclusion. To the gamut of procedures used by JPEG, MPEG adds a blend of motion-compensated, frame-based differential coding (DPCM) techniques. It's actually some-

what misleading to uses the term MPEG as if it refers to a single data
compression method. In fact, MPEG is really a 'toolbox' of compression
techniques that can be applied depending on the difference between net
input data rate and net output data rate required.

Levels and profiles

MPEG-II image compression was defined by the MPEG group in the
International Standards Organization (standards ISO/IEC 13818). MPEG-II
compression used for television is more complex than MPEG-I, which was
originally devised for video compression for desktop video on computers.
However, MPEG-II was written as a super-set of MPEG-I, so MPEG-II
includes MPEG-I compression and adds to it. MPEG-II has four levels that
define the resolution of the picture; from the lowest level (also known as
source input format or SIF), which is the defined input format for MPEG-I
compression, through main level, which represents the standards for
broadcast television pictures, to high level for HDTV pictures.

Low level source input format is derived by down-sampling the
CCIR 601 standard for digital television signals (see Chapter 3). First,
only one interlaced field in every frame is passed to the MPEG encoder.
This not only reduces vertical resolution to half, but also reduces temporal
resolution by the same fraction. To match the vertical resolution loss, the
horizontal resolution is halved too: Luminance samples are reduced from
720 per line to 360 per line (PAL – 625/50), and chrominance resolution is
reduced to 180 pixels per line. Furthermore, chrominance vertical
resolution is halved once more, thus reducing the original picture
resolution by a factor of 16! Main level corresponds to broadcast television
pictures up to a display resolution of 720 active pixels by 576 active lines.
High-1440 level is the high for HDTV pictures up to a resolution of 1440
pixels by 1152 lines, and high level is for wide-screen HDTV resolutions
up to 1920 pixels by 1152 lines.

Profiles relate to the techniques employed by the encoder and
decoder. Simple profile relates to an encoder/decoder pair, which only
uses unidirectional motion estimation and prediction; this saves on coder
and decoder memory requirements but results in a higher bit-rate for a
given overall quality. Main profile is designed for broadcast television
applications; it uses a hierarchy of picture types, which will be described
below, but is more complicated than simple profile because it uses bi-
directional motion vectors, which requires more picture memory. Scaleable profiles are intended for future use, and high profile is intended for
HDTV television pictures. Profiles are hierarchical, so any encoder or
decoder capable of processing, for instance, a main profile can code/
decode a simple profile (a profile downwards), but cannot cope with
scaleable profiles (a profile upwards). Similarly, high profile coders/

decoders can cope with simple and main level, since this is the highest profile.

Main profile at main level

Main profile at main level is a very important combination because it is used for broadcast applications in Europe. Main profile at main level (MP@ML) uses MPEG-II coding of interlaced pictures with resolutions of 720 × 480, 30 Hz frame-rate ('NTSC' pictures) and 720 × 576, 25 Hz frame-rate ('PAL' pictures). MP@ML, MPEG-II coded pictures range in bit-rate from 4 Mbits/s for a quality equivalent to PAL broadcast pictures to 9 Mbits/s, which gives a 'contribution quality' suitable for use in television studio applications (near CCIR-601 quality).

Main level at 4 : 2 : 2 profile (ML@4 : 2 : 2P)

Main level at 4 : 2 : 2 profile has much in common with main profile, except that vertical chrominance bandwidth is maintained once coded in the MPEG domain and a short GOP (see below) is employed to aid editing applications. This profile variant is becoming the *de facto* standard in video production and post-production environments, because 4 : 2 : 2 profile has been found to preserve colour fidelity much better in situations where the video is decoded and encoded several times in the process of television programme production.

Frames or fields

As we have seen, MPEG-I coding dispensed with all the problems caused by interlace by the draconian expedient of taking only one field of an interlaced frame! MPEG-II coding, being a higher level technique and intended to cope with broadcast television applications, cannot take such a cavalier attitude. Instead, MPEG-II allows two-field interlaced pictures to be treated in one of two ways; as frame structure (or progressive), or as field structure. This means that the coder and decoder treat the pictures as full vertical resolution images every 25th of a second (in PAL) or as halved vertical resolution images every 50th of a second. The amount of information to be processed is clearly the same in both instances. Field structure is better at compressing fast movement without artefacts; so it's good for pictures with lots of movement, but it's worse in terms of transform-coding spatial redundancy, so it's worse at compressing still pictures without artefacts. The reverse is true of frame-structure or progressive. Frame/field structure is one of the variables that the broadcaster can choose to alter in order to maximize the quality of transmitted output.

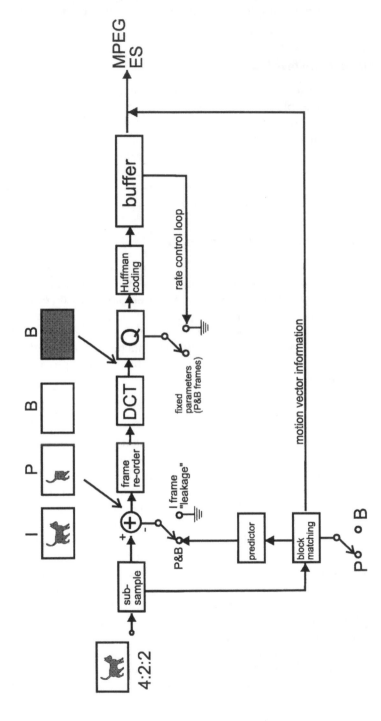

Figure 5.18 *A greatly simplified MPEG coder*

MPEG coding

Figure 5.18 illustrates a greatly simplified MPEG coder operating at main profile and main level. The first circuit block represents the process of sub-sampling the input $4:2:2$, CCIR601 coded television signal to $4:2:0$ format. This process may be as simple as sub-sampling or may involve filtering; implementation issues are not specified in the MPEG standard. In the CCIR601 coded television picture, the chrominance samples are taken half as often as the luminance samples (6.75 MHz sampling for the two chrominance signals against 13 MHz sampling for luminance – see Chapter 3). This is justifiable because the eye is more acute in its perception of luminance information than it is of chrominance. The 601 coding standard (like NTSC) thereby reduces the horizontal bandwidth of colour signals. However, it does nothing to reduce vertical spatial frequencies; unlike SECAM, which takes two lines to send a complete line of colour. The system is therefore wasteful of bandwidth. Clearly, the job of any good data compression system is first to eradicate any over-engineering in the original television signal! Thus, the first stage of MPEG coding is to reduce the vertical spatial frequency of the chrominance samples. The final sampling structure is illustrated in Figure 5.19 and is rather confusingly referred to as $4:2:0$ sampling.

The signal then branches two ways, and here it is worth comparing Figure 5.18 with Figure 5.6; the diagrams for a quantized, error-contained, DPCM compression system. Notice this reflects exactly the structure of the MPEG coder, except that the MPEG system is considerably more complicated. The lower branch shows the block-matching sub-system and predictor. These processes generate the images that are applied to the difference circuit block. Any failures of the predictor simply 'fail' this difference process and are passed on as prediction error for subsequent transform coding.

In MPEG, the level of prediction error is, to some extent, deliberately controlled. Depending on the block-matching algorithm used, three types of picture (frames) are generated in the coding process; these are called P, B

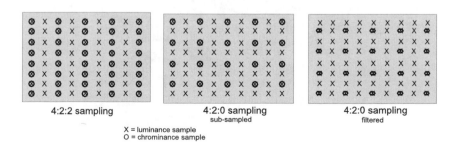

4:2:2 sampling   4:2:0 sampling
sub-sampled

4:2:0 sampling
filtered

X = luminance sample
O = chrominance sample

Figure 5.19   *4 : 2 : 2 and 4 : 2 : 0 sampling structures*

and I frames. P frames are predicted frames and use a forward, unidirectional, block-matching algorithm to generate the movement vectors. This is only a partially effective process because movement in a frame often reveals picture content that was hidden in all previous frames. A simple illustration is given in Figure 5.20, which illustrates a sequence in which a door opens to reveal an external scene. Clearly, the external scene that appears in frame 3 has no precedent in frame 1. (The degree to which the frames differ is greatly exaggerated here for the purposes of illustration.) The prediction error is therefore rather high, because the block-matching system will obviously fail to find an area in the first frame with which to make a match. The difference circuit will, in turn, produce a differential error, which will have to be coded and broadcast. This too is illustrated in Figure 5.20.

Coding efficiency can be improved by using block matching in two directions, as shown. In the case of the door opening, whilst it was not possible to block match the exterior scene in frame 3 with information in frame 1, it is possible to block match the glimpse of the exterior scene in frame 2 with information present in frame 3. Pictures coded like this are known as B frames because the motion vectors are bi-directional or interpolated from adjacent P or I frames. Note how the residual error is nearly zero in the B frame; the following coder stages obviously have a less onerous job to do the less information they receive. Coding efficiency is the great advantage of B frames.

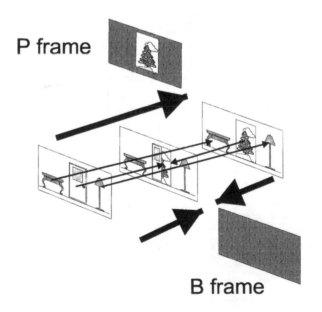

Figure 5.20   *Bi-directional coding and hidden detail*

I frames are a form of leaky DCPM, because the predictor is inter-mittently interrupted and the subsequent coding is thereby fed a complete frame of information with nothing subtracted from it – as illustrated in Figure 5.18. Not only do I frames perform the important function of limiting errors, they also provide essential random-access points in the video bitstream where splices can be performed.

The MPEG coder is programmed to deliver a pattern of I, P and B frames to the ensuing DCT encoder stage. The detail of this pattern is one of a number of variables that may be adjusted for particular applications. For television applications, I, P and B frames are typically assembled into a sequence, termed a group of pictures (a GOP), of 12 frames. This sequence has the structure:

{I B B P B B P B B P B B} {next GOP}

This pattern provides an I frame every 12 frames (at about half-second intervals), which provides a reasonable compromise between coding efficiency and frequency of available splice points.

Taking a closer look at the sequence, the first frame is an I frame. When it leaves the difference circuit, the I frame is a complete video frame. This will be coded downstream by the DCT and quantization blocks. Frame 4 (a P frame) will contain vector information pertaining to movement between this frame and frame 1 (I frame) and any difference information due, for example, to uncovered information, as we saw in Figure 5.20. Frames 2 and 3 will be bi-directionally coded using frames 1 and 4. Now, without some modification, this structure will obviously cause the MPEG decoder a problem because it will try to generate frames 2 and 3, which require the presence of frame 4, before frame 4 has arrived! This is the reason for the circuit block following the difference circuit. This re-orders the frames in the order that the decoder will need to reconstruct the sequence. So

{I B B P ...}

becomes

{I P B B ...}

This process is carried on for the whole GOP like this:

{I P B B P B B P B B I B B}

where the second I is the first I frame of the next GOP.

Following the DPCM part of the MPEG coder, the image sequences pass onto the transform coding stage. This is essentially exactly the same process as described for JPEG compression; including the Huffman coding stage following the transform coding. However, there is one very important further feature of the MPEG coder, and that concerns the feedback loop shown in Figure 5.18.

One of the problems of 'raw' DPCM is the rate variability. If the video sequence contains little movement and no camera movement, the output of the difference stage can be very low indeed. However, if the camera operator performs, for example, a medium rate zoom from a close-up to a wide-shot, the output from the difference circuit can be virtually full-rate video. Fast sequences involving many shot changes and camera movements are particularly troublesome, even with a sophisticated DPCM system with block matching like MPEG. For certain applications this rate variability wouldn't provide a problem, but for television it's clear that the final coding rate must be held steady. To some extent this variability can be managed by a buffer that 'smoothes-out' the overall rate, but eventually any practical (i.e. financially viable) buffer will overflow (or underflow), causing a frame to repeat or be dropped. The resulting image 'hiccup' is very jarring and, whilst this might be an acceptable solution for video conferencing, it is not acceptable for broadcast television. Herein lies the power and the justification for the DCT process in MPEG. Just as JPEG chose to use the DCT so as to allow a broad range of compression ratios, so MPEG uses this wide gamut of possibilities to control the overall coded signal rate.

As we saw with JPEG, it's possible to quantize high-frequency image components much more coarsely than low-frequency elements, and it's even possible to ignore them completely without causing disastrous image degradation. The same is true of MPEG. Should the DPCM stages produce an output stream too fast and too long for the buffer to cope with, the MPEG coder can simply exercise greater quantization of the transform co-efficients, producing at first a soft, fuzzy or, at worst, 'blocky' image, depending on just how drastic an action has to be taken in order to resolve the difference between the overall input–output rate mismatch. (Figure 5.21 illustrates various MPEG-2 artefacts at very low data rates.)

Figure 5.21   *Errors caused when MPEG coding rate is too low*

However, none of these visual effects, even the worst 'blockiness', is as bad as a momentary picture 'jump' or 'freeze'.

To summarize, MPEG uses a block-matching, motion-compensated predictor in a frame-based DPCM loop to reduce temporal redundancy. The predictor is made leaky by transmitting (as often as determined by the user) a frame that is completely not predicted (an I frame). This ensures that errors will not propagate indefinitely, and that there are points in the output bitstream where splices may be performed. The DPCM process yields a series of frames that are either non-predicted (I), predicted from previous frames (P), or predicted from previous and following pictures (B frames). All these frames (suitably re-ordered so as to be in a convenient form for the decoder) then pass to a discrete cosine transform stage, in which the spatial redundancy in the resulting pictures is reduced by quantizing the coefficients of the higher frequency components. The output of the DCT process yields data, which is compressed still further by Huffman coding techniques. The degree to which DCT coefficients need to be quantized is determined by a buffer stage, which regulates the overall output rate. If the output rate of the coder into the buffer starts to rise (as a result of less efficiently coded material) the quantizer stage will start to quantize coefficients more coarsely, thereby maintaining a continuous output data-rate into the following broadcast or distribution chain.

Mosquito noise

Mosquito noise is a colourful term for moving compression-induced errors usually seen around sharp-edged fine text, dense clusters of leaves, and the like. Effectively, mosquito noise is a form of 'ringing' caused by the quantization of the DCT coefficients in the frequency domain. Figure 5.22 ilustrates some 'frozen' mosquito noise.

MPEG coding hardware

Figure 5.23 gives a block diagram of the implementation of the 'DVexpert' 6110 single chip MPEG encoder solution from C-Cube Microsystems of Milpitas, California. C-Cube is a leader in MPEG encoding and decoding technology. The description above of MPEG techniques must have left the correct impression that the development of practical circuitry to implement the techniques involved in encoding MPEG pictures is quite beyond the ability even of medium-scale organizations! In fact, the practical success of MPEG really rests on the availability of highly integrated, powerful silicon solutions from vendors like C-Cube. As the diagram shows, the task for the designer is simply the implementation of interfacing requirements. We shall see the same thing when we come to look at the implementation of MPEG decoding in digital receivers.

Figure 5.22 *Mosquito noise, another MPEG compression artefact*

Statistical multiplexing

The rapid consumer acceptance of digital video is putting pressure on broadcasters continually to introduce new service offerings. Acquiring additional bandwidth to support the new services can be either prohibitively expensive or simply not possible, given, for example, the limited availability of digital terrestrial broadcasting channels. With digital television, extra channels can always be accommodated in a bit-stream by reducing the overall bit-rate of the existing channels and 'shoe-horning' another channel in the overall multiplex. However, there comes a point with this approach where picture quality will deteriorate unacceptably. Statistical multiplexing is a technique that presupposes that not all channels will require the same bandwidth all the time. For instance, channel 1 may have a virtually static scene from a documentary and therefore be producing very little prediction error from the MPEG frame-differential stage for subsequent DCT coding, whilst channel 2 may feature a down-hill skiing competition with fast action and camera movement, producing a great deal of prediction error. Statistical multiplexing (or dynamic bandwidth allocation) enables each channel's 'share' within the multiplex to be changed dynamically – for example, allocating, on a moment by moment basis, a much greater overall data rate to the skiing than to the documentary. In other words, channels with complex scenes borrow bits from channels with less complex scenes. The result is a more efficient use of the total bandwidth, which allows additional channels to be added without any degradation in overall video quality.

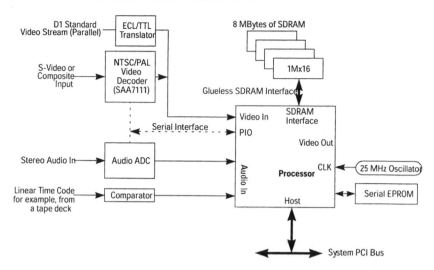

Figure 5.23 *C-Cube 6110 DVexpert MPEG coding chip*

DV, DVCAM and DVCPRO

Just as in MPEG, in the DV digital video standard, the video is sampled at the same rate as 601 digital video, that's to say, 13.5 MHz (720 pixels per scan-line), and the colour information is sampled at half the D-1 rate; although this varies between 525/60 and 625/50 implementations, such that the coding is 4:1:1 in 525-line (NTSC), and 4:2:0 in 625-line (PAL) formats. The sampled video is compressed using a discrete cosine transform (DCT). Unlike MPEG, DV does not utilize motion-compensation and employs only 'intraframe' or frame-by-frame compression. DV video information is carried in a nominal 25 Mbit/s data stream (this stream is sometimes referred to as DV25). To this is added audio, subcode (including timecode), insert and track information (ITI), and error correction, resulting in a final data rate of approximately 36 Mbps.

The frame-by-frame compression specified in the DV standard can cause problems due to the interlaced nature of a frame, because the two merged fields may contain quite different information if the scene has any motion, creating a great deal of false and unnecessary high-frequency information and therefore inflated HF coefficients in the frequency domain. To overcome this problem, the DV standard makes a provision for a new ('2-4-8') DCT 'mode' which effectively reverts to field-based DCT from the normal '8-8' DCT frame-based mode. The DV standard sepcifies how to tell the decompressor which DCT mode is used. DV coding closely resembles MPEG (or more nearly motion-JPEG) coding. That's to say, it utilizes subsampling and DCT transformation. And the similarities don't stop there, because the following processes are quantization of the AC coefficients and entropy coding, just as in JPEG and MPEG. However, the quantization processing is more sophisticated in DV than in MPEG/MJPEG, applying quantization on 'macroblock' level, rather than across a whole frame.

The basic DV (DVC) standard has 'spawned' three important implementations: DV (the consumer, digital, camcorder format), DVCAM, and DVCPRO (professional formats from Sony and Panasonic respectively). The basic video encoding algorithm is the same between all three formats. The consumer-oriented DV uses 10 micron tracks in SP recording mode. Sony's DVCAM professional format increases the track pitch to 15 microns (at the loss of recording time) to improve tape interchange and increase the robustness and reliability of insert editing. Panasonic's DVCPRO increases track pitch and width to 18 microns, and uses a metal particle tape for better durability. DVCPRO also adds a longitudinal analog audio cue track and a control track to improve editing performance and user-friendliness in linear editing operations.

In a variation of DV, Sony's Ditigal8 uses digital recording with DV compression but bases the tape format on the existing Video8/Hi8 technology. Digital8 records on Video8 or Hi8 tapes, but these run at twice their normal speed. Digital8 will also playback existing Video8 and Hi8 tapes.

DV, being a frame-based compression scheme, is ideal for video editing applications. Many computer-based video editing packages use DV. Interfacing products between the world of DV and CCIR 601 professional television equipment include products such as the DV-Bridge from Miranda Technologies, shown in Figure 5.24.

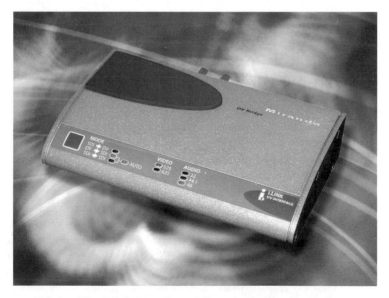

Figure 5.24   *The DV-Bridge from Miranda Technologies Ltd*

6
Audio data compression

In principle, the audio engineering problem presented by low-data bandwidth, and therefore in reduced digital resolution, is no different to the age-old analogue problem of restricted dynamic range. In analogue systems, noise reduction systems (like Dolby B or dbx1) have been used for many years to enhance the dynamic range of inherently noisy transmission systems like analogue tape. All of these analogue systems rely on a method called 'compansion', a word derived from the contraction of compression and expansion, which refers to the dynamic range being deliberately reduced (compressed) in the recording stage processing and recovered (expanded) in the playback electronics. In some systems this compansion acts over the whole frequency range (dbx is one such type). Others work over a selected frequency range (Dolby A, B, C and SR). We shall see that the principle of compansion applies in just the same way to digital systems of data reduction. Furthermore, the distinction made between systems that act across the whole audio frequency spectrum and those that act selectively on ranges of frequencies (sub-bands) is also true of digital implementations. However, the digital systems chosen for digital television have carried the principle of sub-band working to a degree of sophistication undreamed of in analogue implementations.

Compression based on logarithmic representation

Consider the 8-bit digital values 00001101, 00011010, 00110100, 01101000 and 11010000 (8-bit examples are used because the process is easier to follow, but the principles below apply in just the same way to digital audio samples of 16 bits or, indeed, any word length). We might just as correctly write these values thus:

$$00001101 = 1101 \times 1$$

$$00011010 = 1101 \times 10$$

$$00110100 = 1101 \times 100$$

$$01101000 = 1101 \times 1000$$

$$11010000 = 1101 \times 10000$$

If you think of the multipliers 1, 10, 100 and so on as powers of two, then it's pretty easy to appreciate that the representation above is a logarithmic description (to the log of base two) with a 4-bit mantissa and a 3-bit exponent. So already we've saved 1 bit in 8 (a 20 per cent data reduction). We've paid a price, of course, because we've sacrificed accuracy in the larger values by truncating the mantissas to 4 bits. These crude techniques work even in practical (but non-critical) audio applications like telecoms because of the phenomenon of masking, which underlies the operation of all noise reduction systems (see Chapter 2). Put at its simplest, masking is the reason we strain to listen to a conversation on a busy street and why we cannot hear the clock ticking when the television set is on: Loud sounds mask quiet ones. So the logarithmic representation makes sense because resolution is maintained at low levels but sacrificed at high levels, where the programme signal will mask the resulting, relatively small, quantization errors.

NICAM

Further audio data-rate reductions may be made because real audio signals do not change instantaneously from very large to very small values, so the exponent value may be sent less often than the mantissas (even if they did, we wouldn't hear it due to the effect of temporal masking). This is the principle behind the stereo television technique of NICAM, which stands for near instantaneous companded audio multiplex. In NICAM 728, 14-bit samples are converted to 10-bit mantissas in blocks of 32 samples with a common 3-bit exponent. This is an excellent and straightforward technique, but it is only possible to secure relatively small reductions in data throughput of around 30 per cent.

Psychoacoustic masking systems

Wide-band compansion systems view the phenomenon of masking very simply; and rely simply on the fact that programme material will mask system noise. But masking is actually a more complex phenomenon. Essentially, it operates in frequency bands and is related to the way in which the human ear performs a mechanical Fourier analysis of the incoming acoustic signal. It turns out that a loud sound only masks a quieter one when the louder sound is lower in frequency than the quieter, and only then when both signals are relatively close in frequency (see Chapter 2). It is due to this effect that all wide-band compansion systems can only achieve relatively small gains. The more data we want to discard, the more subtle our data reduction algorithm must be in its appreciation of the human masking phenomenon. These compression systems are termed psychoacoustic systems and, as you will see, some systems are very subtle indeed.

MPEG layer I compression (PASC)

It's not stretching the truth too much to say that the failed Philips' digital compact cassette (DCC) system was the first non-professional digital audio tape format. Until DCC, digital audio developments had ridden on the back of video technology. The CD rose from the ashes of Philips' laserdisc and DAT machines, using the spinning-head tape recording technique originally developed for B- and C-format one-inch video machines and later exploited in U-Matic and domestic videotape recorders. It is to their credit then that, in developing the digital compact cassette, Philips chose not to follow so many other manufacturers down the route of modified video technology. Inside a DCC machine there was no head-wrap, no spinning head and few moving precision parts. Until DCC, it had taken a medium suitable for recording the complex signal of a colour television picture to store the sheer amount of information needed for a high-quality digital audio signal. Philips' remarkable technological breakthrough in squeezing two high-quality stereo digital audio channels into a final data rate of 384 kBaud was accomplished by, quite simply, dispensing with the majority (75 per cent) of the digital audio data! Philips named their technique of bit-rate reduction, or data-rate compression, precision adaptive sub-band coding (PASC). PASC was adopted as the original audio compression scheme for MPEG video/audio coding (layer I).

In MPEG (layer I) or PASC audio coding, the whole audio band is divided up into 32 frequency sub-bands by means of a digital wave filter. The filters are relatively simple, and provide good time resolution with reasonable frequency resolution. At first sight it might seem that this process would increase the amount of data to be handled tremendously – or by 32 times anyway! This, in fact, is not the case, because the output of the filter bank for any one frequency band is at 1/32nd of the original sampling rate. If this sounds counter-intuitive, take another look at the Fourier transform in Chapter 4, where the details of the discrete Fourier transform are given, and note that a very similar process is being performed here. Observe that, when a periodic waveform is sampled n times and transformed, the result is composed of n frequency components. Imagine computing the transform over a 32-sample period: Thirty-two separate calculations will yield 32 values. In other words, the data rate is the same in the frequency domain as it is in the time domain. Actually, considering both describe exactly the same thing with exactly the same degree of accuracy, this shouldn't be surprising. Once split into sub-bands, sample values are expressed in terms of a mantissa and exponent, exactly as explained above. The layer I algorithm groups together 12 samples from each of the 32 bands. The maximum magnitude in each block is used to establish the masking 'profile' at any one moment and thus predict the mantissa accuracy to which the samples in that sub-band can be reduced without their quantization errors becoming perceivable (see Figure 6.1).

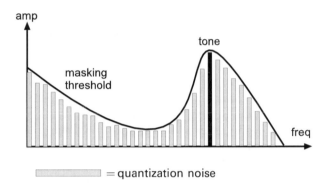

Figure 6.1   *Sub-band quantization and how it relates to masking profile*

MPEG layer II audio coding (MUSICAM)

The MPEG layer II algorithm is the preferred algorithm for European DTV, and includes a number of simple enhancements of layer I. Layer II was originally adopted as the transmission coding standard for the European digital radio project (digital audio broadcasting or DAB), where it was termed MUSICAM. The full range of bit-rates for each layer is supported, as are all three sampling frequencies, 32, 44.1 and 48 kHz. Note that MPEG decoders are always backward-compatible; for example, a layer II decoder can decode layer I or layer II bitstreams, but cannot decode a layer III encoded stream.

MPEG layer II coding improves compression performance by coding data in larger groups. The Layer II encoder forms frames of $3 \times 12 \times 32 = 1152$ samples per audio channel. Layer I codes data in single groups of 12 samples for each sub-band, whereas layer II codes data in three groups of 12 samples for each sub-band. The encoder encodes with a unique scale factor for each group of 12 samples only if necessary to avoid audible distortion. The encoder shares scale factor values between two or all three groups when the values of the scale factors are sufficiently close, or when the encoder anticipates that temporal noise masking will hide the consequent distortion (see Chapter 2). The layer II algorithm also improves performance over layer I by representing the bit allocation, the scale factor values and the quantized samples with a more efficient code. Furthermore, layer II coding added 5.1 multi-channel capability. This was done in a scaleable way, so as to be compatible with layer I audio. An improvement to the standard with respect to Pro Logic compatibility led to a second edition of the MPEG layer II standard, accepted in 1997. These adaptations are discussed in the next chapter.

MPEG layers I and II contain a number of engineering compromises. The most severe concerns the 32 constant-width sub-bands, which do not accurately reflect the equivalent filters in the human hearing system (the critical bands). Specifically, the bandwidth is too wide for the lower frequencies, so the number of quantizer bits cannot be specifically tuned for the noise sensitivity within each critical band. Furthermore, the filters have insufficient Q, so that signal at a single frequency can affect two adjacent filter bank outputs. Another limitation concerns the time–frequency–time domain transformations achieved with the wave filter. These are not transparent so, even without quantization, the inverse transformation would not perfectly recover the original input signal.

MPEG layer III

The layer III algorithm is a much more refined approach. Layer III is finding its application on the Internet, where the ability to compress audio files by a large factor is important in download times. It is not used in digital television applications.

Dolby AC-3

The analogy between data compression systems and noise reduction has already been drawn. It should therefore come as no surprise that one of the leading players in audio data compression should be Dolby, with that company's unrivalled track record in noise reduction systems for analogue magnetic tape. Dolby AC-3 is the adopted coding standard for terrestrial digital television in the US; however, it was actually implemented for the cinema first, where it was called Dolby Digital. It was developed to provide multi-channel digital sound with 35 mm prints. In order to retain an analogue track so that these prints could play in any cinema, it was decided to place the new digital optical track between the sprocket holes, as illustrated in Figure 6.2. The physical space limitation (rather than crude bandwidth) was thereby a key factor in defining its maximum practical bit-rate. Dolby Labs did a great deal of work to find a channel format that

Figure 6.2   *Dolby Digital as originally coded in film stock*

would best satisfy the requirements of theatrical film presentation. They discovered that five discrete channels, left (L), right (R), centre (C), left surround (LS) and right surround (RS), set the right balance between realism and profligacy! To this they added a limited (1/10th) bandwidth sub-woofer channel, the resulting system being termed 5.1 channels. Dolby Digital provided Dolby Labs with a unique springboard for consumer formats for the new DTV (ATSC) systems.

Like MPEG, AC-3 divides the audio spectrum of each channel into narrow frequency bands of different sizes optimized with respect to the frequency selectivity of human hearing. This makes it possible to sharply filter coding noise so that it is forced to stay very close in frequency to the frequency components of the audio signal being coded. By reducing or eliminating coding noise wherever there are no audio signals to mask it, the sound quality of the original signal can be subjectively preserved. In this key respect, a perceptual coding system like AC-3 is essentially a form of very selective and powerful Dolby-A type noise reduction! Typical final data-rate applications include 384 kb/s for 5.1-channel Dolby surround digital consumer formats, and 192 kb/s for two-channel audio distribution.

Jo Coleman

Information Update Service

Butterworth-Heinemann

FREEPOST SCE 5435

Oxford

Oxon

OX2 8BR

UK

Keep up-to-date with the latest books in your field.

Visit our website and register now for our FREE e-mail update service, or join our mailing list and enter our monthly prize draw to win £100 worth of books. Just complete the form below and return it to us now! (FREEPOST if you are based in the UK)

www.bh.com

Please Complete In Block Capitals

Title of book you have purchased:...

....

Subject area of interest:...

Name:...

Job title:...

Business sector (if relevant):...

Street:..

Town:.. County:...

Country:... Postcode:...

Email:...

Telephone:..

How would you prefer to be contacted: Post ☐   e-mail ☐   Both ☐

Signature:.. Date:...

☐ Please arrange for me to be kept informed of other books and information services on this and related subjects (✔ box if not required). This information is being collected on behalf of Reed Elsevier plc group and may be used to supply information about products by companies within the group.

FOR OFFICE USE ONLY

Butterworth-Heinemann,
a division of Reed Educational
& Professional Publishing Limited.
Registered office: 25 Victoria Street,
London SW1H 0EX.
Registered in England 3099304.
VAT number GB: 663 3472 30.

BUTTERWORTH
HEINEMANN

A member of the Reed Elsevier plc group

7
Digital audio production

Digital line-up levels and metering

Suppose I asked you to put together a device comprising component parts I had previous organized from different sources, and that I had paid very little attention to whether each of the component parts would fit together (perhaps one part might be imperial and another metric). You would become frustrated pretty quickly, because the task would be impossible. So it would be too for the audio engineer if digital signals were not, to some degree at least, standardized. The rational behind these standards and the tools used in achieving this degree of standardization are the subjects of the first few sections of this chapter.

The adoption of standardized studio levels (and of their associated line-up tones) ensures the interconnectability of different equipment from different manufacturers and that tapes made in one studio are suitable for replay and/or rework in another. Unfortunately these standards have evolved over many (analogue!) years, and some organizations have made different decisions; this, in turn, has reflected upon their choice of operating level. National and industrial frontiers exist too, so that the subject of maximum and alignment signal levels is fraught with complications.

Fundamentally, there are only two absolute levels in any digital system; maximum level and quantization noise floor. These are both illustrated in Figure 7.1. Any signal that is lower than the noise floor will disappear as it's swamped by noise, and signal that is larger than maximum level will be distorted. All well recorded signals have to sit comfortably between the 'devil' of distortion and the 'deep blue sea' of noise. Actually, that's the fundamental job of any television sound-engineer!

In principle, maximum level would make a good line-up level. Unfortunately, it would also reproduce over loudspeakers as a very loud noise indeed, and would therefore, likely as not, fray the nerves of

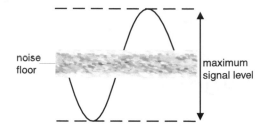

Figure 7.1   *The dynamic range of an audio system*

those people working day after day in studios! Instead a lower level is used for line-up, which actually has no physical justification at all but is cleverly designed to relate maximum signal level to the perceptual mechanism of human hearing and to human sight (!) as we shall see. Why sight? Because it really isn't practical to monitor the loudness of an audio signal by sound alone. Apart from anything else, human beings are very bad at making this type of subjective judgement. Instead, from the very earliest days of sound engineering, visual indicators have been used to indicate audio level; thereby relieving the operator of making subjective auditory decisions. There are two important and distinct reasons to monitor audio level. The first is to optimize the drive, the gain or sensitivity of a particular audio circuit, so that the signal passing through it is at a level whereby it enjoys the full dynamic range available from the circuit. If a signal travels through a digital circuit at too low a level, it unnecessarily picks up quantization-noise in the process. If it is too high, it may be distorted or 'clipped', as shown in Figure 7.1.

The second role for audio metering exists in radio or television continuity studio, where various audio sources are brought together for mixing and switching. Viewers and listeners are justifiably entitled to expect a reasonably consistent level, and do not expect one programme to be unbearably loud (or soft) in relation to the last. In this case, audio metering is used to assess the apparent loudness of a signal and thereby make the appropriate judgements as to whether the next contribution should be reduced (or increased) in level compared with the present signal. The two operational requirements described above demand different criteria of the meter itself. This pressure has led to the evolution of two types of signal monitoring meter; the volume unit (VU) meter and the peak programme meter (PPM).

The VU meter

A standard VU meter is illustrated in Figure 7.2. The VU is a unit intended to express the level of a complex wave in terms of decibels above or

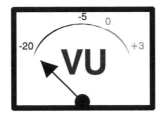

Figure 7.2   *Standard VU meter calibration*

below a reference volume; it implies a complex wave – a programme
waveform with high peaks. 0 VU reference level therefore refers to a
complex-wave power-reading on a standard VU meter. The VU meter is
an indicator of the average power of a waveform; it therefore accurately
represents the apparent loudness of a signal because the ear, too,
mathematically integrates audio waveforms with respect to time. However, because of this, the VU is not a peak-reading instrument. A failure to
appreciate this, and on a practical level this means allowing the meter
needle to swing into the red section on transients, means the mixer is
operating with inadequate system headroom. This characteristic has led
the VU to be regarded with suspicion in some organizations.

To some extent this is unjustified, because the VU may be used to
monitor peak levels provided the action of the device is properly understood. The usual convention is to assume that the peaks of the complex
wave will be 10–14 dB higher than the peak value of a sine wave adjusted
to give the same reference reading on the VU meter. In other words, if a
music or speech signal is adjusted to give a reading of 0 VU on a VU
meter, the system must have at least 14 dB headroom – over the level of a
sine wave adjusted to give the same reading – if the system is not to clip
the programme audio signal. In operation, the meter needles should only
very occasionally swing above the 0 VU reference level on a complex
programme.

The PPM meter

Whereas the VU meter reflects the perceptual mechanism of the human
hearing system, and thereby indicates the loudness of a signal, the PPM is
designed to indicate the value of peaks of an audio waveform. It has its
own powerful champions, notably the BBC and other European broadcasting institutions. The PPM is suited to applications in which the balance
engineer is setting levels to optimize a signal level to suit the dynamic
range available from a transmission channel. Hence its adoption by
broadcasters, who are under statutory regulation to control the depth of
their modulation and therefore fastidiously to control their maximum

signal peaks. In this type of application, the balance engineer does not need to know the 'loudness' of the signal, but rather needs to know the maximum excursion (the peak value) of the signal.

It is actually not difficult to achieve a peak reading instrument; the main limitation of this approach lies with the ballistics of the meter itself which, unless standardized, leads to different readings. The PPM standard demands a defined and consistent physical response time of the meter movement. Unfortunately, the simple arrangement is actually unsuitable as a volume monitor due to the highly variable nature of the peak-to-average ratio of real-world audio waveforms; a ratio known as crest-factor. This enormous ratio causes the meter needle to flail about to such an extent that it is difficult to interpret anything meaningful at all! For this reason, the PPM meter itself is preceded by a logarithmic look-up table. This effectively compresses the dynamic range of the digital signal prior to its display; a modification that greatly enhances the PPM's readability.

The peak programme meter of the type used by the BBC is illustrated in Figure 7.3. Notice the scale marked 1 to 7, each increment representing 4 dB (except between 1 and 2, which represents 6 dB). This constant deflection per decade is realized by the logarithmic stage. Line-up tone is set to PPM4 and signals are balanced so that peaks reach PPM6; that is, 8 dB above reference level. (BBC practice is that peak studio level is 8 dB above alignment level.) BBC research has shown that the true peaks are actually about 3 dB higher than those indicated on a BBC PPM, and that operator errors cause the signal occasionally to swing 3 dB above the indicated permitted maximum; i.e. a total of 14 dB above alignment level.

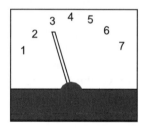

Figure 7.3   *Standard BBC type peak programme meter (PPM)*

Opto-electronic level indication

Electronic level indicators range from professional bargraph displays, which are designed to mimic VU or PPM alignments and ballistics, through the various peak reading displays common on consumer and 'prosumer' digital goods (often bewilderingly calibrated) to simple peak-indicating LEDs. The latter can actually work surprisingly well – and facilitate a degree of precision alignment which belies their extreme simplicity.

In fact, the difference between monitoring using VUs and PPMs is not as clear-cut as stated. Really, both meters reflect a difference in emphasis: The VU meter indicates loudness – leaving the operator to allow for peaks based on the known, probabilistic nature of real audio signals. The PPM, on the other hand, indicates peak – leaving it to the operator to base decisions of apparent level on the known stochastic nature of audio waveforms. However, the latter presents a complication because, although the PPM may be used to judge level, it does take experience. This is because the crest factor of some types of programme material differs markedly from others, especially when different levels of compression are used between different contributions. To allow for this, institutions that employ the PPM apply *ad hoc* rules to ensure continuity of level between contributions and/or programme segments. For instance, it is BBC practice to balance different programme material to peak at different levels on a standard PPM.

Despite its powerful European opponents, a standard VU meter combined with a peak sensing LED is very hard to beat as a monitoring device, because it both indicates volume and, by default, average crest factor. Any waveforms that have an unusually high peak-to-average ratio are indicated by the illumination of the peak LED. PPMs unfortunately do not indicate loudness, and their widespread adoption in broadcasting accounts for the many uncomfortable level mismatches between different contributions – especially between programmes and adverts.

Standard operating levels and line-up tones

Irrespective of the type of meter employed, it should be pretty obvious that a meter is entirely useless unless it is calibrated in relation to a particular signal level (think about if rulers had different centimetres marked on them!).

Three important line-up levels exist:

- PPM4 = 0 dBu = 0.775 V RMS, used by UK broadcasters
- 0 VU = +4 dBu = 1.23 V RMS, used in the commercial music sector
- 0 VU = −10 dBV = 316 mV RMS, used in consumer and 'prosumer' equipment.

Digital line-up

The question of how to relate 0 VU and PPM 4 to digital maximum level of 0 dBFS (0 dB relative to full scale) has been the topic of hot debate. Fortunately, the situation has crystallized over the last few years to the extent that it is now possible to describe the situation on the basis of

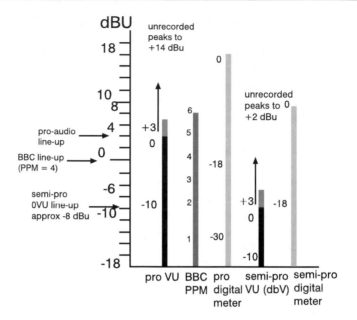

Figure 7.4 *Digital and analogue programme and alignment levels*

widespread implementation in USA and European broadcasters. Essentially,

$$0\,\text{VU} = +4\,\text{dBu} = -20\,\text{dBFS (SMPTE RP155)}$$

$$\text{PPM4} = 0\,\text{dBu} = -18\,\text{dBFS (EBU R64-1992)}$$

Sadly these recommendations are not consistent, and, whilst the EBU recommendation seems a little pessimistic in allowing an extra 4 dB headroom above their own worst-case scenario, the SMPTE suggestion looks positively gloomy in allowing 20 dB above alignment level. This probably reflects the widespread – though technically incorrect – methodology, when monitoring with VUs, of setting levels so that peaks often drive the meter well into the red section. Figure 7.4 illustrates the relationship between various line-up levels in the analogue and digital domain. (It should also be noted that, despite the EBU recommendation, implementations vary. For instance, in Germany it is still common to refer 0 dBFS to + 15 dBu, and to + 22 dBu in France!)

Switching and combining audio signals

The simplest form of AES router is effectively a RS422 router with transformer coupling. This is termed an asynchronous AES switcher. However, in order to be able to switch silently between digital audio sources (i.e. without an audible 'splat') the AES router must perform switch transitions at data word-boundaries. For that to happen, all the digital signals must be co-timed within the switcher frame, irrespective of their various timings at each AES signal input. The operation performed in the process of re-timing each input is termed re-framing. Essentially, each signal is fed into a FIFO (first-in-first-out) buffer and is read out at the appropriate time so as to be synchronous with the AES reference signal supplied to the switcher frame. A routing switcher of this type is known as a synchronous router, and is necessarily considerably more complicated and expensive than the simple, asynchronous type.

Digital audio consoles

In television audio production, it is usually the case that each individual contributor is separately miked and the ensemble sound is mixed electrically. It is the job of the balance engineer to control this process. This involves many aesthetic judgements; however, there are relatively few parameters under the balance engineer's control. Over and above the office of correctly setting the input sensitivity control so as to ensure best signal to noise ratio and control of channel equalization, the balance engineer's main duty is to judge and adjust each channel gain fader and, therefore, each contributor's level within the mix. A further duty, when performing a stereo mix, is the construction of a stereo picture or image by controlling the relative contribution each input channel makes to the two stereo outputs. The apparent position of each contributor within the stereo picture (image) is controlled by a special stereophonic panoramic potentiometer, or pan pot for short.

Sound mixer architecture

The simplest form of digital audio mixer is illustrated in Figure 7.5. In this case, two digital audio signals are each multiplied by coefficients (k1 and k2) derived from the position of a pair of fader controls, one fader

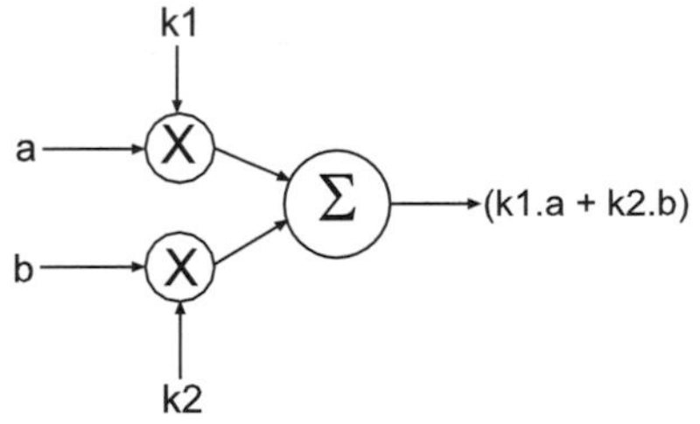

Figure 7.5 *A simple digital mixer*

assigned to either signal. The signals issuing from these multiplication stages are subsequently added together in a summing stage. All audio mixers possess this essential architecture supplemented many times over.

Mixer automation

Mixer automation consists (at its most basic level) of computer control over the individual channel faders. Even the most dextrous and clear thinking balance engineer obviously has problems when controlling perhaps as many as 24 or even 48 channel faders at once. For mixer automation to work, several things must happen. First, the controlling computer must know precisely which point in the TV programme has been reached, in order that it can implement the appropriate fader movements. Secondly, the controlling computer must have, at its behest, hardware that is able to control the audio level on each mixer channel swiftly and noiselessly. A third requirement of a fader automation system is that the faders must be 'readable' by the controlling computer so that the required fader movements can be implemented by the human operator and memorized by the computer for subsequent recall.

A complete fader automation system is shown in schematic form in Figure 7.6. Notice that the fader simply acts as a potentiometer driven by a stabilized supply. By digitizing the control voltage and making this value available to the microprocessor bus, the fader 'position' can be stored for later recall. When this happens, digital value is re-created and this is

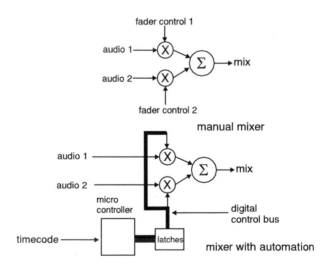

Figure 7.6   *An automated digital audio mixer under the control of time-code*

applied to the digital multiplier, thereby reproducing the operator's original intentions.

One disadvantage of this type of system is the lack of operator feedback. Importantly, when in recall·mode the faders fail, by virtue of their physical position, to tell the operator (at a glance) the condition of any of the channels and their relative levels. Some automation systems attempt to emulate this important visual feedback by creating an iconic representation of the mixer on the computer screen. Some even allow these virtual faders to be moved, on screen, by dragging them with a mouse. Another more drastic solution is to use motorized faders, with the control system loop reading and 're-creating' operator fader physical movements, thereby providing the quite thrilling spectacle of banks of faders moving as if under the aegis of ghostly hands! Note that timecode is the means by which an automation system is kept in step with other processes within the television production.

Digital tape machines

Digital tape recorders may be distinguished by their basic technology pedigree. One type uses stationary heads in the manner of an analogue tape recorder, the so-called digital audio stationary head (DASH) format. The other uses rotating heads, in the manner of videotape machines (see Table 7.1).

Digital two-track recording

The cassette based digital audio tape (DAT) two-track tape format, originally pioneered by Sony but now very widely standardized, is now virtually universal as a professional mastering format. The DASH method produced mastering formats that permitted the mechanical editing of digital tape, an attribute that was considered important in the early days of digital recording. However, with the advent of hard-disk editing this requirement is no longer required. DAT technology uses a small videocassette style tape cassette and a small rotating head assembly. The wrap angle on the tape is, however, very much smaller than that used in most video formats, and this has a number of advantages. It reduces tape drag and wear (which makes the format more suitable for portable applications), and it makes the threading process less complicated than that of a videocassette machine. This simplification is possible because, although the bandwidth required for digital audio is indeed greater than that required for its analogue counterpart, it is not as high as that required for analogue video. Furthermore, because the signal is digital the signal-to-noise requirement is much less stringent, so Sony took advantage of various simplifications that may be made to the video style mechanism

7.1   *Digital recording formats*

Tracks	Head	Format	Medium/speed	Application
2	Stationary	DCC	DCC tape	High-quality replay
2	Rotary	1610/1630	60FPS Umatic	Editing/CD mastering
2	Rotary	PCM-F1	NTSC Betamax	Semi-pro recording
2	Rotary	DAT	DAT tape	High-quality mastering
2	–	MiniDisc	MiniDisc	High-quality replay
2	–	Hard disk	Winchester Disk	Editing/CD mastering
2–4	Stationary	DASH	1/4″ 7.5 ips	High-quality mastering
4–16	Stationary	DASH	1/4″ 30 ips	High-quality multi-track
8	Rotary	ADAT	S-VHS	High-quality multi-track
8	Rotary	DA-88	Hi-8 mm video-tape	High-quality multi-track
24–48	Stationary	DASH	1/2″ 30 ips	High-quality multi-track

when designing this bespoke digital-audio tape format. DAT achieves a remarkably high data capacity, typically consuming 8.15 mm of tape per second: nearly six times slower than the linear tape speed in a standard analogue cassette!

Digital multi-tracks

The DASH recorders use multi-track head assemblies to write multiple tracks of digital audio onto tape. With the advent of low-cost rotary-head multi-tracks, DASH is becoming extinct. Figure 7.7 is an illustration of the multi-track nature of the data format on a DASH machine.

The two most common formats for digital multi-track work are the rotary-head videocassette-based Alesis ADAT family (Figure 7.8) and the Tascam DA-family. The ADAT format records on readily available

individual tracks

Figure 7.7 *Arrangement of DASH tracks on tape*

Figure 7.8 *ADAT rotary digital multi-track machines*

S-VHS videocassettes, and up to 16 ADAT recorders can be linked together for 128 tracks with no external synchronizer required and without sacrificing a track to timecode. The most common alternatives to the ADAT family are the DA-88 (or DA-38) by Tascam, which use Hi-Band 8 mm compact cassettes (Hi-8 mm) instead of the S-VHS tapes chosen by Alesis.

Digital audio workstations

When applied to a digital audio application, a computer hardware platform is termed a digital audio workstation or DAW. Editing digital audio on a desktop microcomputer has two major advantages:

1 An edit may be made with sample accuracy – that is, a cut may be made with a precision of 1/40 000th of a second!
2 An edit may be made non-destructively – that is, when the computer is instructed to join two separate takes together, it doesn't create a new file with a join at the specified point but instead records two pointers which instruct it (on subsequent playback) to vector or jump to another data location and play from the new file at that point.

In other words, the computer 'lists' the edits in a new file of stored vector instructions. Indeed this file is known as an edit decision list. (Remember that the hard disk doesn't have to jump instantaneously to another location because the computer holds a few seconds of audio data in a RAM cache memory.) This opens the possibility of almost limitless editing in order to assemble a 'perfect' performance. Edits may be rehearsed and auditioned many times without ever 'molesting' the original sound files.

Audio file formats

Digital representations of sound, when on computer, are stored just like any other kind of data; as files. There are a number of different file formats in common usage. Most sound files begin with a header consisting of information describing the format of that file. Characteristics such as word length, number of channels and sampling frequency are specified so that audio applications can properly read the file. One very common type of file format is the WAV (or Wave) format. This is a good example because it demonstrates all the typical features of a typical audio file.

WAV files

WAV files are a version of the generic RIFF file format that was co-developed by Microsoft and IBM. RIFF represents information in pre-defined blocks, preceded by a header that identifies exactly what the data are. This format is very similar to the AIFF format developed by Apple (see below) in that it supports monaural and multi-channel samples and a variety of sample rates. Like AIFF, WAV files are big and require approximately 10 Mbytes per minute of 16-bit stereo samples with a sampling rate of 44.1 kHz. Here is a hexadecimal representation of the first 128 bytes of a WAV file:

```
26B7:0100  52 49 46 46  28 3E 00  00-57 41 56  45 66 6D 74  20 RIFF(> ...WAVEfmt
26B7:0110  10 00 00 00  01 00 01  00-22 56 00  00 22 56 00  00 ..."V..."V...
26B7:0120  01 00 08 00  64 61 74  61-04 3E 00  00 80 80 80  80 ...data.>...
26B7:0130  80 80 80 80  80 80 80  80-80 80 80  80 80 80 80  80 ........
26B7:0140  80 80 80 80  80 80 80  80-80 80 80  80 80 80 80  80 ........
26B7:0150  80 80 80 80  80 80 80  80-80 80 80  80 80 80 80  80 ........
26B7:0160  80 80 80 80  80 80 80  80-80 80 80  80 80 80 80  80 ........
26B7:0170  80 80 80 80  80 80 80  80-80 80 80  80 80 80 80  80 ........
```

The header provides Windows™ with all the information it needs. First off, it defines the type of RIFF file; in this case, WAVEfmt. Note the bytes which are shown underlined. The first two, 22 and 56, relate to the audio sampling frequency. Their order needs reversing to read 5622 hexadecimal, which is equivalent to 22050 in decimal – in other words,

22 kHz sampling. The next two inform the file player software that the sound file is 1 byte per sample (mono) the following, 8 bits per sample.

AU files

AU (or u-law – pronounced mu-law) files utilize an international standard for compressing audio data. It has a compression ration of $2:1$. The compression technique is optimized for speech (in the United States it is a standard compression technique for telephone systems; in Europe, a-law is used). This file format is most frequently found on the Internet, where it is used for '.au' file formats, alternately know as 'Sun audio' or 'NeXT' format. Even though it's not the highest quality audio file format available, its non-linear logarithmic coding scheme results in a relatively small file size; ideal for applications where download time is a problem.

AIFF and AIFC

The audio interchange file format (AIFF) allows for the storage of monaural and multi-channel sample sounds at a variety of sample rates. AIFF format is frequently found in high-end audio recording applications. Originally developed by Apple, this format is used predominantly by Silicon Graphics and Macintosh applications. Like WAV, AIFF files can be quite large; one minute of 16-bit stereo audio sampled at 44.1 kHz usually takes up about 10 Mbytes. To allow for compressed audio data, Apple introduced the new AIFF-C, or AIFC, format, which allows for the storage of compressed and uncompressed audio data. AIFC supports compression ratios as high as 6:1. Most of the applications that support AIFF playback also support AIFC.

MPEG

As well as its presence in digital television, the International Standard Organization's Moving Picture Expert Group is responsible for one of the most popular compression standards in use on the Internet today. Designed for both audio and video file compression, MPEG audio compression specifies three layers, and each layer specifies its own format. The more complex layers take longer to encode but produce higher compression ratios while keeping much of an audio file's original fidelity. Layer I takes the least amount of time to compress, but layer III yields higher compression ratios for comparable quality files, as we saw in the last chapter.

VOC

Creative voice (.voc) is the proprietary sound file format that is recorded with Creative Lab's Sound Blaster and Sound Blaster Pro audio cards. This format supports only 8-bit mono audio files up to sampling rates of 44.1 kHz and stereo files up to 22 kHz.

Raw PCM data

Raw pulse code modulated data is sometimes identified with the .pcm, but at times it has no extension at all. Since no header information is provided in the file, you must specify the waveform's sample rate, resolution and number of channels to the application to which it is loaded.

Surround-sound formats

Dolby Surround

Walt Disney Studio's *Fantasia* was the first film ever to be shown with a stereo soundtrack. That was in 1941. Stereo in the home has been a reality since the 1950s. Half a century on, it's reasonable that people might be looking for 'something more'. With the advent of videocassette players, watching film at home has become a way of life. Dolby Surround was originally developed as a way of bringing part of the cinema experience to the home, where a similar system named Dolby Stereo has been in use since 1976. Like Dolby Stereo, Dolby Surround is essentially a four-channel audio system encoded or matrixed into the standard two stereo channels. Because these four discrete channels are encoded within the stereo channels, extra hardware is required both at the production house and in the home. Decoders are now very widespread because of the take-up of analogue technology-based home cinema systems, and digital television is expected to continue – even accelerate – interest in this market. The extra hardware required, in addition to normal stereo, consists of a number of extra loudspeakers (ideally three), a decoder and an extra stereo power amplifier. Some manufacturers supply the decoder and four power amplifiers in one AV amplifier unit. In addition, a sub-woofer channel may be added (a sub-woofer is a loudspeaker unit devoted to handling nothing but the very lowest audio frequencies – say below 100 Hz). Frequencies in this range do add a disproportionate level of realism to reproduced sound. In view of the very small amount of information (bandwidth), this is surprising. However, it is likely that humans infer the scale of an acoustic environment from these subsonic cues.

A typical surround listening set up is illustrated in Figure 7.9. Note the extra two channels, centre and surround, and the terminology for the final

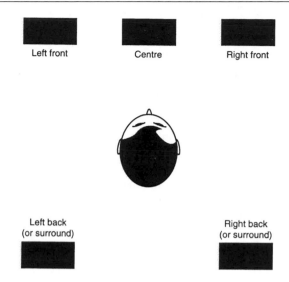

Figure 7.9   *A typical surround-sound set-up*

matrixed two channel signals Lt and Rt, standing for left-total and right-total respectively. The simplest form of decoder (which most certainly does not conform to Dolby's criteria, but is nevertheless reasonably effective) is to feed the centre channel power amplifier with a sum signal (Lt + Rt) and the surround channel amplifier with a difference signal (Lt − Rt). This bare-bones decoder works because it complements (to a first approximation) the way a Dolby Surround encoder matrixes the four channels onto the left and right channels: centre channel split between left and right, surround channel split between left and right with one channel phase reversed. If we label the original left/right signals L and R, we can state the fundamental process formally:

Input channels:
Left (sometimes called left music channel)   L
Right (sometimes called right music channel)   R
Centre channel (sometimes called dialogue channel)   C
Surround channel (for carrying atmosphere sound effects etc.) S

Output channels (encoding process):

$$Lt = i(L + jC + kS)$$

$$Rt = i(R + jC - kS)$$

where i, j and k are simply constants. And the decoding process yields:

Left $(L') = e(Lt)$
Right $(R') = f(Rt)$
Centre $(C') = u(Lt + Rt) = u[i(L + jC + kS + R + jC - kS)] = u[i(L + R + 2jC)]$
Surround $(S') = v(Lt - Rt) = v[i(L + jC + kS - R - jC + kS)] = v[i(L - R + 2kS)]$

where e and f and u and v are constants.

This demonstrates that this is far from a perfect encoding and decoding process. However, a number of important requirements are fulfilled even by this most simple of matrixing systems, and to some extent the failure mechanisms are masked by operational standards of film production. Dolby have cleverly modified this basic system to ameliorate the perceptible disturbance of these unwanted crosstalk signals. Looking at the system as a whole – as an encode and decode process – first, and most importantly, note that no original centre channel (C) appears in the decoded rear, surround signal (S'). Also note that no original surround signal (S) appears in the decoded centre channel (C'). This requirement is important because of the way these channels are used in movie production. The centre channel (C) is always reserved for mono dialogue. This may strike you as unusual, but it is absolutely standard in cinema audio production. Left (L) and right (R) channels usually carry music score. Surround (S) carries sound effects and ambience. Therefore, considering the crosstalk artefacts, at least no dialogue will appear in the rear channel – an effect that would be most odd! Similarly, although centre channel information (C) crosstalks into left and right speaker channels (L' and R'), this only serves to reinforce the centre dialogue channel. The most troublesome crosstalk artefact is the $v(iL - iR)$ term in the S' signal, which is the part of the left/right music mix that feeds into the decoded surround channel – especially if the mix contains widely panned material (with a high interchannel intensity ratio). Something really has to be done about this artefact for the system to work adequately, and this is the most important modification to the simple matrix process stated above that is implemented inside all Dolby Surround decoders. All decoders delay the S' signal by around 20 ms which, due to an effect known as the law of the first wavefront or the Hass effect, ensures that the ear and brain tend to ignore the directional information contained within signals that correlate strongly with signals received from another direction but at an earlier time. This is certainly an evolutionary adaptation to avoid directional confusion in reverberant conditions, and biases the listener, in these circumstances, to ignore unwanted crosstalk artefacts. This advantage is further enhanced by band-limiting the surround channel to around 7 kHz and using a small degree of high-frequency expansion. Dolby Pro Logic enhances the system still more by controlling the constants written as e, f, u and v above dynamically, based on programme information. This technique is known as adaptive matrixing. Because Dolby Surround is a matrixed system, it is fundamentally compatible with stereo sound systems (like NICAM 728). However, the DTV specifications have been more ambitious still in recommending true multi-channel sound for digital television.

Dolby digital (AC-3)

As we saw in the last chapter, Dolby AC-3 is the adopted coding standard for terrestrial digital television in the US. However, it was actually implemented for the cinema first, where it was called Dolby Digital. Dolby Surround Digital or Dolby Surround AC-3 provides separate channels for left, right, and centre speakers at the front; two surround speakers at the sides; and a sub-woofer at the listener's option. When multiple two-channel AES digital inputs are used, the preferred channel assignment is:

Pair 1: left, right
Pair 2: centre, LFE
Pair 3: left surround, right surround

Unlike the analogue Dolby Surround, with its single band-limited surround channel (usually played over two speakers – see above), Dolby Digital features two completely independent surround channels, each offering the same full range fidelity as the three front channels.

Rematrixing

When the AC-3 coder is operating in a two-channel stereo mode, an additional processing step is inserted in order to enhance interoperability with Dolby Surround 4-2-4 matrix encoded programs. The extra step is referred to as 'rematrixing', whereby the signal spectrum is broken into four distinct rematrixing frequency bands. Within each band the energy of the left, right, sum and difference signals are determined. If the largest signal energy is in the left or right channel, the band is encoded normally. If the dominant signal energy is in the sum or difference channel, then those channels are encoded instead of the left and right channels. The decision as to whether to encode left and right or sum and difference is made on a band-by-band basis, and is signalled to the decoder in the encoded bitstream.

Dynamic range compression

It is common practice for high quality programming to be produced with wide dynamic range audio, suitable for the highest quality audio reproduction environment. Broadcasters, serving a wide audience, typically process audio in order to reduce its dynamic range. The AC-3 audio coding system provides an embedded dynamic range control system that allows a common encoded bitstream to deliver programming with a dynamic range appropriate for each individual listener. A dynamic range control value is provided in each audio block (every 5 ms), and these values are used by the audio decoder in order to alter the level of the reproduced audio for each audio block. Level variations of up to 24 dB may be indicated.

MPEG-II extension to multi-channel audio

The ITU-R Task Group TG10-1 has worked on a recommendation for multi-channel sound systems. The main outcome of this work is Recommendation BS.775, which says that a suitable multi-channel sound configuration should contain the 5.1 channels already discussed in relation to AC-3. The extension to multi-channel sound supports up to five input channels, the low frequency enhancement channel and up to seven commentary channels in one bitstream.

Pro logic compatibility

When the source material is already surround-encoded (e.g. Dolby Surround), the broadcaster may choose to transmit this directly to the audience in stereo only mode (i.e. two-channel). This appears to be the normal situation in Europe, where any move to 5.1 is so slow as to be unnoticeable! Compatibility with existing surround decoders is assured by several means. The multi-channel encoder can operate using a surround-compatible matrix. This will allow stereo decoders to receive the surround-encoded signal, with optional application to a surround decoder. A full multi-channel decoder will rematrix all the signals to obtain the original multi-channel presentation. This mode is supported in the MPEG-II multi-channel syntax and, consequently, in the DVB specification.

IEC 61937 interface

As yet, most digital receivers in Europe do not have a built-in digital multi-channel decoder (because interested users have often already got a home-cinema set-up); instead, they feature nothing more than stereo analogue audio at $0\,VU = -10\,dBV$ level on phono sockets. This implies that any downstream Pro Logic, AC-3 or MPEG multi-channel decoding (most of which is digital) will require another analogue-to-digital conversion stage before decoding. However, some decoders have an output interface for the coded MPEG-II audio multi-channel bitstream for connection to an external decoder. Such an interface is defined as IEC standard IEC 61937.

This is essentially the SPDIF interface described in Chapter 3, with encoded data instead of the usual linearly coded audio payload, and the validity bit set to 'non-valid' to indicate the signal should not be converted directly to analogue using a DAC. (If you try this it gives a really horrible 'buzz'!) The data capacity of the SPDIF interface is adequate for transmission of coded bitstreams. Data is transmitted in bursts. The distance between the start of consequent bursts corresponds to the MPEG audio frame length of 1152 PCM samples. The length of each burst corresponds to the bit rate, e.g. 24×384 bits when the bit-rate equals 384 kbit/s at 48 kHz sampling frequency. Each burst is preceded by a preamble, which provides information on the length of the burst.

Dynamic range compression

This issue was already discussed in relation to the AC-3 coding used by ATSC. MPEG sound coding, chosen by DVB, allows the implementation of such a system by providing an ancillary data field, within which the compression information may be transmitted. Alternatively, the ISO/MPEG Layer I and II standard offers an even more attractive solution. By adjusting all scale factors (see Chapter 6) with a single gain factor, the compression can be performed digitally (and automatically) in the decoder. The steps in gain that occur at the 8-ms block boundaries are effectively smoothed by the windowing action of the sub-band filter bank – it's effectively NICAM with the compression taken out!

Multilingual support

MPEG audio supports two options for multilingual:

1 Separate audio streams for each language. This option is very flexible, but is not so bandwidth efficient as the alternative below.
2 Embedding up to seven language channels in a specially defined multi-channel audio stream. Such an audio stream may, for example, contain a normal multi-channel programme at 384 kbit/s, plus a number of reduced bandwidth voice channels each occupying an extra 64 kbit/s.

Editing MPEG layer II audio

Near-seamless editing is possible in the coded domain with a resolution of 24 ms. Due to the characteristics of the sub-band filter as applied in MPEG audio layer II, a short cross-fade will automatically be generated at the editing points. This avoids clicks in the audio. The only problem is that 24 ms doesn't divide into television field frequency, so the problem becomes where to cut the picture and where to cut the sound?

8
Digital video production

Swi4tching and combining video signals

A television programme consists of a sequence of different images, as the 'eye' of the camera flicks (cuts) from viewpoint to viewpoint. Stylistically this production technique originated in film, where sections of film (shots or sequences) are joined together – literally – with glue. In television this effect is achieved electronically by switching between the output from one camera and that of another. For this to be accomplished successfully, the switch must occur during the vertical interval; this is directly analogous with the mechanical joining of film, which must be executed at a frame boundary for the cut to be invisible. This sounds innocuous enough, but it actually places requirements both on the switch itself and, more importantly, on the video sources to be selected. Crucially, the vertical interval of both pictures must happen at the same time – or be in synchronism. This is most important, and indicates that any video source contributing to a system (digital or analogue) must be both synchronous and co-timed. After all, it's no use switching in the vertical interval of one signal to another during active picture time. Not only will the cut be visible as a flash, but the final decoded TRS sync information may well have an 'extra' field sync pulse, which will cause a monitor to roll as it loses vertical sync. This requirement is normally achieved by feeding timing reference information (usually called colour-black) to each video source so that it produces signals in time with the reference. When they are so arranged, the video sources are said to be genlocked to the colour-black reference. (Note that, even in a fully digital television studio, the reference is often still analogue colour-black.)

Digital television signals appear in two main forms, as discussed in Chapter 3. The 10-bit pulse code modulation (PCM) television signals are either in a bit parallel or bit serial form. In bit parallel form, the signals are in the form of 74 ns symbols which describe the luminance and chrominance of each pixel in terms of 8- or 10-bit binary numbers. At this rate, the data is invariably handled internally within equipment as 3-, 8- or 10-bit TTL logic signals and their associated clocks; one signal for luminance and two for chrominance. Source switching is performed using standard fast TTL (F-series) or advance CMOS (AC-series) multiplexing techniques (Figure 8.1).

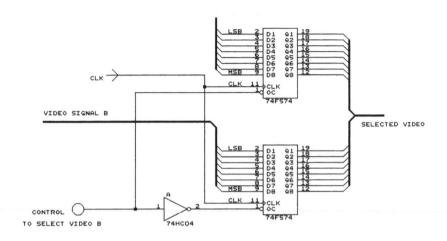

(a)

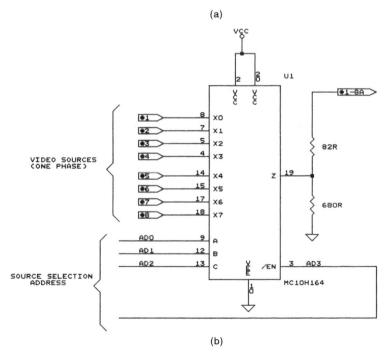

(b)

Figure 8.1 (a) *Parallel digital video switching circuit;* (b) *Serial digital video switching circuit (in ECL logic)*

This is the form in which most digital video data is handled for video processing. Digital television signals in serial form are, in some ways, more akin to their analogue precursors. The symbol rate in this case is very high; up to 360 Mbits/s! Unlike parallel digital signals they flow in one circuit, which must, due to the high symbol rate, be a matched termination line. Serial video switching is usually performed by emitter-coupled-logic (ECL) elements, which are the only digital IC family with switching times fast enough to cope with this very high data rate. Parallel digital video signals are the norm within studio equipment, and serial video signals are the norm for digital interconnection. Typical interconnections look like Figure 8.2. In the process of parallel-to-serial conversion, considerable steps are taken to remove low-frequency content caused by a long stream of symbols in the same state (1s or 0s). Unfortunately some remain, and this requires the serial digital signal be DC restored prior to being re-sliced by the input circuit.

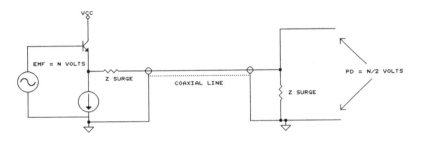

Figure 8.2   *A transmission line*

Digital video effects

As we shall see, video effects require the offices of signal multiplication devices. Digital multiplication can be achieved in many ways, but a good technique involves the use of a look-up-table (LUT). Essentially, multiplier and multiplicand are used together to address one unique reference in a read-only-memory (ROM) where the result can be looked up, as in Figure 8.3.

What is a video transition?

A television programme comprising only one video shot would be pretty boring, no matter how complex the shot! That's why vision-switchers were invented; so that programme-makers could select between shots. The process of transferring between two video shots is known as a video transition. Sometimes – when the process isn't live – the process itself is called editing, and its practitioners, editors. Nowadays, dedicated vision

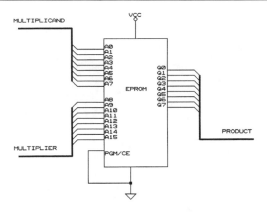

Figure 8.3 *Digital multiplier using an EPROM*

switchers are being replaced by video editing software running on PCs and workstations. But however the technology achieves its result, essentially there are only four types of video edit or transition:

1 The cut
2 The mix or dissolve
3 The fade
4 The wipe.

Each of these video transitions has a different visual 'meaning' and connotation. The vision switcher or the video editing software running on a PC or workstation possesses the ability to combine different editing techniques together, and this enables variety and drama to be added to video productions.

The cut

The cut is the most common form of video transition or edit. In a cut, one video shot is replaced instantaneously by another. The process is illustrated in Figures 8.4 and 8.5. Technically the cut is a simple switch and, as we have seen, the action of the switch must be synchronized to the video and occur during the vertical interval when no picture information is present. The cut is the simplest form of video edit. There are three others; the dissolve, the wipe and the fade.

The dissolve

The dissolve (sometimes referred to as a mix, a lap-dissolve or just a lap) is a transition where one video shot is gradually replaced by another. Specifically, the first shot gradually gets fainter whereas the second shot

Figure 8.4   *In a cut one shot cuts . . .*

Figure 8.5   *to another!*

starts invisible and gradually gets more visible. Halfway through, each shot contributes equally to the final picture. Figure 8.6 illustrates the mix transition. In it, picture A dissolves into picture B. Technically, the dissolve is achieved by multiplying all the image pixels in one video source by one coefficient and all the pixels in the other image source by one minus that coefficient and adding the results together. So if the two video sources are thought of as P and Q, and the final output is $aP + (1 - a)Q$, a dissolve is achieved as a is varied from 1 (where the final output is all P) to 0 (where the output is all Q). In software digital video editing programs these

Figure 8.6   *A mix*

calculations are obviously achieved within the software itself. In hardware
vision mixers, hardware multiplication circuits are employed.

The fade

The fade can be divided into two types; the fade-in and the fade-out. The
fade-out is the gradual transition from an image to a black screen, whereas
the fade-in is the gradual transition from a black screen to an image. These
transitions are typically used by professionals at the end and beginnings of
programmes respectively, or to show the passage of time. As you may
have guessed already, the fade can be thought of as a dissolve or a mix but
– instead of to another shot – to a black signal. Technically the fade is the
equivalent of the dissolve, except that one video input is replaced by a
black signal.

Wipes

A wipe transition involves one video shot being gradually replaced by
another. It's a bit like a mix in that the transition isn't instantaneous as in a
cut but, unlike a mix, in a wipe one picture is gradually revealed as a
pattern moves across the screen – 'wiping' the old picture out and revealing
the new picture beneath. Figure 8.7 illustrates a diagonal wipe – so called
because of the diagonal pattern. Wipes are generated by means of line and
field rate counters. These generate linear functions with respect to time. If
these signals are fed to a digital comparator and a variable signal is fed to its
other input, as the variable input is gradually changed the transition of the

Figure 8.7　*A wipe effect*

comparator will change at various times during the line and field scan. Because this happens on every line, if this signal is fed as signal *a* to the dissolve circuit mentioned above where picture P and picture Q are multiplied like this:

$$aP + (1 - a)Q = \text{output signal}$$

the result is an abrupt transition from one signal to another, which is variable according to the pre-set value on one terminal of the comparator. More complicated patterns can be derived by the summation and non-additive mixing of line rate and field rate digital counts.

Split-screens

This effect (in effect a 'frozen' wipe) is used by professional vision mixers in news production programmes, where news readers have a still from the news sitting above their shoulder as they read the introduction to the story. Figure 8.8 is an illustration.

Keys

Essentially, luminance keying is a process in which a negative hole is cut electronically in the original picture and video from the caption source is used to fill the 'key-hole'. Figure 8.9 illustrates the process. You can think of keying to a first approximation as a very high bandwidth switch. In high quality broadcast mixers, this 'switch' is actually performed by two complimentary multipliers so that the switch is performed with a controlled rise-time. Because the keying is controlled by multiplication, graphics can be superimposed with degrees of transparency.

Figure 8.8   *Split-screen*

Figure 8.9   *A luminance key*

Posterize

Posterizing is a very dramatic video effect. You see it on pop videos and other video productions that feature extensive use of special effects. The effect of posterizing is illustrated in Figure 8.10. The effect is simply quantization distortion, and is achieved by curtailing the number of bits of digital video information to less than the usual 8 (or 10) bits assigned to luminance data.

Figure 8.10   *Posterize effect*

Chroma-key

The chroma-key is one of the most powerful ways that live-action video can be combined with computer generated backdrops or animations. Figure 8.11 illustrates the power of the effects obtainable from this feature; the teddy bear is in a studio, not on a cliff-top! The chroma-key signal itself is a picture modulation signal, derived from picture chrominance information, which is used to matte two television images together. A particular colour is chosen as a keying colour; blue is often used because it is very

Figure 8.11   *A chroma-key*

different from the colour of flesh tones. (This ensures the minimum possibility of a false-key, where part of the subject appears transparent and the background image emerges as if through some anatomically impossible aperture!) However, other colours are employed; bright green and yellow, for instance – again chosen for their dissimilarity to the hues of human skin. Operationally, a blue chroma-key is set up so that the foreground image (shot in front of an all blue set) is processed and a signal derived that is used to suppress the blue areas of the foreground picture, and the inverse of this signal is employed to modulate the other picture over the areas of blue and 'fill' these with an artificial background image.

The first stage of the keying process involves isolating a particular hue from within the foreground image. Note the term hue; the task required is the isolation of a particular wavelength (actually a range of wavelengths) of light, irrespective of its intensity. This is a very important distinction, because shadows are inevitably cast within the blue set – however carefully it has been lit to avoid this. If the shadows are not to destroy the key, the process must continue to derive a constant key in the face of different luminance levels. Fortunately, in television, there already exist signals that describe the chrominance of a scene irrespective of the luminance of the colours involved. These are the colour-difference signals; $B - Y$ or Cb and $R - Y$ or Cr, which were described in Chapters 2 and 3. The simplest method of deriving a blue chroma-key involves passing the Cb $(B - Y)$ signal to a comparator circuit or digital 'clipper', as this type of circuit is usually called. This circuit produces a unipolar switching signal that can be used to select the appropriate areas of foreground and background image. Figure 8.12 illustrates this process. Unhappily, this simple arrangement has a number of operational disadvantages. First, the key may only be blue, and secondly, the switching nature of the circuit produces a distinctive 'fizz' on keyed edges.

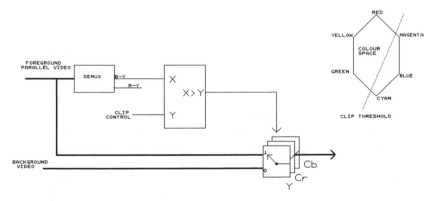

Figure 8.12 *A simple digital chroma-key circuit arrangement*

Keying colour may be made infinitely variable by multiplying the
(B − Y) and (R − Y) signals by sine and cosine functions and adding
the products; a technique that produces two new signals, sometimes
termed i and q. This mathematical manipulation is, in effect, a rotation of
the colour plane, and is equivalent to the 2D image rotation illustrated
later in Figure 8.22. The clipper circuit thus acts on the i signal as it did on
the Cb (B − Y) signal in Figure 8.12, but this time, the colour plane itself is
rotated as appropriate. A further enhancement is gained if the q signal
(which is in quadrature) is passed to an absolute value processor. This
secondary signal may now be subtracted from the i signal in variable
proportions to vary the angle of colour selectivity.

A third disadvantage of the simple chroma-key circuit noted above is
due to the effect of lighting 'spill' from the coloured flat (or set), which
reflects and falls upon the foreground subject. Typically, flesh tones take
on a cadaverous, blue tinge, which only serves to augment the unnatural
look of a chroma-key matte. Colour-correction techniques (described in
Chapter 4) may be employed to ameliorate this effect, but a more
complete solution, and one which reduces the edge 'fizz' so often
associated with simple chroma-keys, involves the use of subtractive or
linear (rather than multiplicative or switched) keying. In a subtractive
chroma-key, the key-colour signal in the original foreground image is
isolated and subtracted from the original image. Areas that were originally
blue are thus rendered black, and are thereby made ready to have an
alternative image fill these black areas. This is achieved by multiplying the
background image by an inverted version of the isolated key-colour
signal. In the process, the blue spill light is subtracted from the foreground
image, thus restoring a more natural colour balance. Modern chroma-key
units (perhaps the most famous are due to Ultimatte) are so refined that a
modern subtractive chroma-key may be impossible to discern, even to the
trained eye.

Off-line editing

An important role for the computer in video editing, aside from its duties
as image processor (as if that weren't enough!), is the control of video
sources and logging of edit decisions in relation to the timecode values
associated with the original material. It's worth noting that some video
techniques are very difficult to achieve in real-time on a desktop computer. Often, the image quality may leave something to be desired
(depending on the amount and type of compression employed). Nevertheless, the computer is able to produce a list of all the edits – in relation to
timecode values – their duration, the video sources involved etc. and
produce this as a listing known as an edit-decision list (EDL). This EDL
may be kept in machine readable form and taken to a much more

expensive broadcast-quality on-line video editing suite, where the process originally carried out on the desktop system can be repeated using top-flight broadcast equipment much faster than if the editor had gone to the on-line suite with nothing but the videotapes and ideas! This process is referred to as off-line editing (to distinguish it from on-line), and it still represents one duty of desktop video editing systems in the broadcast video environment.

Computer video standards

So far we have only considered the television video standards of NTSC and PAL, and some of the new possible HDTV formats. The salient details of these systems were tabulated in Chapters 2 and 5. It's important to realize that the choices of numbers of lines and field rates used in these systems were just engineering choices made long ago. Most people now don't remember the British 405-line television system or the French 819-line system, but their existence proves that there's nothing set in stone about 525 or 625 lines. Perhaps, unfortunately, no one needed to tell this to the designers of computer video systems! As a result, they have produced a number of different standards; the most important of which are considered below.

The video sub-system of a PC is illustrated in Figure 8.13. The system shown is a typical SVGA configuration. Especially important is the direct analogue connection of red, green and blue drives from video digital to analogue converter to monitor. Computer video systems do not bother to encode the colour signal. Really there's no need to do this, as bandwidth is not a problem and the signal remains much 'cleaner' due to the absence of cross-colour artefacts. The video card sub-system consists of a video

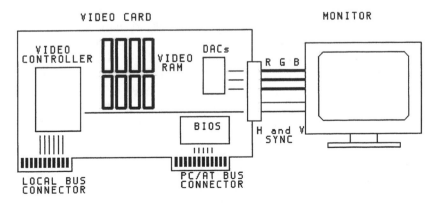

Figure 8.13   *PC video sub-system*

controller chip, video RAM memory, BIOS and DACs. The video chip does all the hard work converting the image held in video memory (VRAM) into data, which are in a suitable form to send to the DACs to be displayed as video. It is also responsible for the creation of monitor synchronizing information and general 'housekeeping' duties. Crudely put, the amount of video RAM determines the resolution and number of colours obtainable from the video system. Some cards have as little as 0.5 Mbytes, some as much as 4 Mbytes. The BIOS chip handles the communications between the PC's CPU and the video sub-system through the ISA and local bus connectors or the PCI (peripheral components interconnect) bus on newer machines.

VRAM is mapped as part of the PC's memory which both the CPU and the video card BIOS can address simultaneously. An almost perfect television image of a real scene may be obtained if each primary colour has a signal-to-noise ratio of around 50 dB. In digital terms, this requires three 8-bit values for each pixel of the image. In software this is known as a TrueColor image, and there are 16.8 million unique combinations (two raised to the power of 24; sometimes simply referred to as '16 million colours') which may be displayed in a TrueColor image. The video RAM in the PC video system may well be as much as 2 Mbytes for 800×600 SVGA. This is so that it can store and display $800 \times 600 \times 3 = 1.4$ Mbytes necessary to define a photographic-quality image. Unfortunately IBM, when they first designed the PC, allowed for a 64k window in the address space of the CPU, which may be used to address video memory! A big problem, because the address space is less than the available memory. This hardware limitation requires that software addressing schemes known as bank-switching must be employed, which shift the addressing window to different parts of the video memory; they do this by allocating some of the data bits as address bits.

The majority of desktop applications intended for business users will not handle TrueColor images, and instead make do with a very reduced number of colours. In the most common PC display modes (SVGA 800×600, 256 colour for example), 1 byte is used to describe each pixel, so only 256 accessible colours are available in any frame. This sounds like an almost crippling restriction, and it would be were it not for the technique known as colour palettes. In a paletted image, a pixel value isn't representative of a colour but a reference to a palette table. The data file must therefore contain not only the 2D array, which is the bitmap itself, but also palette information.

Figure 8.13 also illustrates the sync information being fed to the monitor from the PC. Most computer video standards have abandoned interlaced scanning, the norm in the television world. Furthermore, it is common in the computer world to keep vertical-sync and horizontal-sync signals separate. Most monitors accept sync information in the form of quasi-TTL

Table 8.1 *Computer scanning standards. The close relationship between NTSC and VGA can be seen clearly*

Standard	Resolution	H freq (kHz)	Total lines	Active lines	V freq	S
VGA	640 × 480	31.469525	480	59.94	(−)	
VESA SVGA	800 × 600	48.077666	600	72.19	(+)	
VESA SVGA	1024 × 768	56.476806	768	70.07	(−)	
Mac 2 Page	1152 × 870	68.681915	870	75.06	(−)	

signals (2–5 V). Some standards have negative-going syncs, pulse low to initiate retrace; some pulse high. The most common PC video scanning and sync standards and the Mac II Two-Page standard are compared in Table 8.1.

AppleColor and Apple Monochrome monitors have a 640 × 480 graphics capability, and refresh at 60.01 Hz or 66.67 Hz. Twenty-four-bit colour boards are available for photo-realistic image processing on NuBus equipped Macs.

Vector and bitmap graphics – what's the difference?

Vector-based and raster-based or bitmap-based graphics systems use different internal representations for the images they reproduce. We have already met bitmap images in the form of the raster images of analogue and digital television signals. In a bitmap representation, the image is simply an addressable 2D array of numbers where the values contained in the array identify the colour (or luminance, in the case of grey-scale images) for the corresponding area of the image. A vector-based system stores graphics objects as sets of primitives; lines of certain length, direction and width, curves, colour fills etc., as well as instructions of how they are to be arranged and – importantly – in what order. Vector graphics are sometimes referred to as object-oriented graphics, and they are suited to very high-speed applications since geometrical translations, rotations and so on are easy to re-compute. A bitmap image, on the other hand, is very memory- and computation-intensive. True vector display systems do exist in some high-end optically coupled systems (the old phrase for virtual reality systems) used for military use. In all PC based applications, however, bitmap displays are the norm: All the computer video standards mentioned above are raster display systems. That's not to say that vector graphics have no role to play; it's just that the vector approach exists only in software – the vector-based representation is always converted to a bitmap for the PC to display it. TrueType fonts represent an extremely important application of vector graphics-based techniques.

Graphic file formats

Table 8.2 lists the most common graphics file formats you'll come across working on desktop microcomputers and workstations in television applications.

Now let's look in detail at some of the more important file formats. PICT is the standard Mac graphics file format. Virtually any Mac program will allow the export and import of PICT files. Windows bitmap (.BMP/ .DIB) is probably the most common picture file on the PC. PCX is also a PC-based file, one of the oldest, and developed originally by ZSoft. GIF and JPEG files are considered in detail, since these are device-independent standards. TARGA (.TGA) is a high-end standard used for high-end graphics.

Table 8.2   *The most common graphics file formats in television applications*

Image format	File extension (IBM PC)	Type of file
Windows Bitmap	.BMP	Bitmap
Drawing Exchange file	.DXF	Vector used by Auto-CAD
Encapsulated Post-Script	.EPS	Vector
GEM file	.GEM	Bitmap
Graphics Interchange Format	.GIF	CompuServe bitmap
HPGL	.PLT	HP plotter – vector
JPEG	.JPG	Compressed bitmap
PhotoCD	.PCD	Bitmap
PICT	.PCT, PICT	Mac bitmap standard
PCX	.PCX	Oldest bitmap format
Lotus Pic	.PIC	Vector
Tagged Image File	.TIF	Bitmap
Windows Metafile	.WMF	Bitmap

Windows Bitmap (.BMP)

Almost always, .BMP images are 256 or 16 colour. The palette entry which resides in the image header is 4 bytes long for each entry and contains RGB components (the fourth byte being reserved). Interestingly, the image data is written from the bottom to the top (the opposite of a television raster) in this format.

PCX

One of the oldest bitmap formats, PCX is extensively used for image interchange in the IBM PC environment. This file format is the one chosen by Corel PhotoPaint. PCX format allows a number of different options and pixel value depths up to 24 bit where there is no palette specified. Images may be compressed or uncompressed.

TARGA

TARGA is a high-end graphics file format. Video depth can go to 32 bit; RGB values and a key signal. TARGA files hardly ever specify a palette. Images may be compressed or uncompressed. (The compression technique employed is run-length encoding, which is explained in Chapter 5.)

GIF

GIF (pronounced jiff) was developed by CompuServe as a machine-independent file format. Because programs quickly developed that allowed GIF images to be viewed on the PC and the Mac, it wasn't long before subscribers begun taking this format beyond its original role of on-line graphics for CompuServe and started using it as a handy file format for swapping graphics from different types of computers. Video depth may extend to 8 bit per pixel, so it is a low-end application. However, it has very wide support and is almost certainly the best 8-bit paletted graphics format. All images are compressed using a LZW compression algorithm, which is very effective. LZW is a compression technique based on the coding of repeated data chains or patterns. Effectively it does for image binary data what a palette does for colours – it sets up a table of common patterns and codes specific instances of patterns in terms of 'pointers', which refer to much longer sequences in the table. The algorithm doesn't use a pre-defined set of patterns, but instead builds up a table of patterns which it 'sees' from the incoming data. (LZW compression is described in Chapter 5.) Note that the algorithm does not look for patterns within the image itself; only in the resulting data. Compression algorithms that analyse the image itself are available, and JPEG is the most important amongst these. If we use the DOS Debug utility to examine a .GIF file, the 'top' of it looks like this;

```
26B7:0100  47  49  46  38  37  61  92  01-2E  01  80  00  00  00  00  00 GIF87a ...
26B7:0110  FF  FF  FF  2C  00  00  00  00-92  01  2E  01  00  02  FF  8C ......
26B7:0120  8F  A9  CB  ED  0F  A3  9C  B4-DA  8B  B3  DE  BC  FB  0F  86 ......
26B7:0130  E2  48  96  E6  89  A6  EA  CA-B6  EE  0B  C7  F2  4C  D7  F6 H ... L ...
26B7:0140  8D  E7  FA  CE  F7  FE  0F  0C-0A  87  C4  A2  F1  88  4C  2A ...... L*
26B7:0150  97  CC  A6  F3  09  8D  4A  A7-D4  AA  F5  8A  CD  6A  B7  DC ... J j ...
26B7:0160  AE  F7  0B  0E  8B  C7  E4  B2-F9  8C  4E  AB  D7  EC  B6  FB ... N ...
26B7:0170  0D  8F  CB  37  80  B9  FD  5E-AC  E3  F7  FC  9D  BE  0F  18 ... 7 ...^ ...
```

Note the 6-byte header; GIF87a. This specifies the data stream as GIF and that this particular file is the 87a version. On the second line, the HEX value 2C is the standard value for the first byte in the Image descriptor section; following this is the origin for the left and top position of the image (both zero in this case), then 4 bytes, two of which denote image width (0192 Hex) and the following two image height (012E Hex). If the GIF file contains palette information, it immediately follows this image descriptor section. The data follows last.

JPEG

JPEG was a file format before it became a part of television and still is! In fact, JPEG has vast application in the field of photographically acquired images. More than a file format, it also defines a special lossy compression algorithm, as we saw in Chapter 5.

Computer generated images (CGI) and animation

The role of the computer in graphics and animation is very wide indeed. It may be used solely as a drawing tool, to create individual, powerful images – in other words, in a graphics role. Alternatively, it may be used to create and store a succession of individual images and sequence them together into the final animated clip. If it's a powerful computer it may be able to do this in real-time; otherwise it may have to dump each frame to tape, or disk, in an intermittent (single-frame) process. It may be used to create tweens, where the machine creates 'in-between' images, averaging the motion between individual drawn frames, to give a more continuous, flowing sense of movement. Finally, the computer may generate an artificial three-dimensional world where complex three-dimensional moves, synthetic light sources, shadows and surface textures conspire in a process known as rendering to produce individual images quite beyond those achievable by human artists without several lifetimes at their disposal. (One lifetime, at least, to perform the maths!)

There are two basic types of animation and two basic types of graphics. Together they may be thought of as four components within a matrix of possibilities derived from different approaches in software and different hardware abilities.

Animation

	Real-time	Non-real-time (single-frame)
2D Graphics	*	*
3D Graphics	*	*

Types of animation

In Chapter 2, attention was drawn to an important attribute of the human eye that has notable relevance to the technology of film and video. This property, the persistence of vision, refers to the phenomenon whereby an instantaneous cessation of light does not result in a similarly instantaneous discontinuation of signals within the optic nerve and visual processing centres. Consequently, if the eye is presented with a succession of slightly different still images at a adequately rapid rate, the impression is gained of a moving image. A movie film camera operates by capturing real, continuous movement in a sequence of temporally sampled still images, which are reconstructed by the 'low-pass filter' of our visual processing. Animation techniques take the illusion one step further. Here the process begins with a sequence of still images which, when viewed swiftly enough, give the impression of a moving image. To see an animator at work is a truly wonderful process; the verb to animate literally means 'to breath life into'.

The emphasis on what follows concerns animation performed on desktop personal computers. Fortunately, newer generation desktop computers are equipped with excellent graphics quality which rivals or betters broadcast television resolution. However, there are many issues concerning the ability (or lack of it) simply to store images generated on various computers on other mediums (especially videotape). Most of these compatibility problems concern the different line and field standards in use between the computer and television industries. Real-time animation systems have the ability to replay individual frames at the overall frame-rate required for them to be viewed as animation. This obviously places considerable requirements on hardware, since not only must the machine be equipped with a great deal of RAM and a fast processor but also normal disk access times may ultimately be too slow to keep up with the absolute sustained data-rate required.

In a single-frame system, the animator makes the decision how good he or she wants the output quality to be. The computer doesn't dictate this. The artist takes all the time needed to generate the image and dumps the images intermittently onto the storage medium. Single-frame animation enables fairly humble machines to create excellent results – but, of course, it does require playing out from a medium other than the desktop machine. Ultimately these distinctions will disappear. It's only hardware limitations that prevent real-time animation on all kinds of computers. One day, real-time photographic quality animations will be commonplace on personal computers.

Software

The simplest kind of computer animation program exploits a computer's ability to generate and store images. In this approach the computer is used to draw and colour individual frames, and employs typical drawing tools such as scaling, rotation and cut and paste – combined with textual tools, mattes and so on – to generate a series of frames suitable for animation. Each frame, once finished, is stored on hard disk or videotape for later application. Animation techniques of this type are known as 2D systems. Such a medium is the planar world of the fine artist, rather than the technician, because it is up to the artist to provide the invented world's sense and rendering of depth, including shading and shadows. These techniques are really computerized versions of the old hand-drawn animation techniques. A 3D animation program generates a sense of depth by calculation of perspective, shadows, shading and so on. The animator's role may therefore be more conceptual than artistic, and involve the specification of components and their trajectories within a spatial frame. In contrast with 2D techniques, the computer does the bulk of the imaging in 3D systems.

2D systems

Draw and paint functions form the basis of 2D graphics and animation systems. In addition to the usual drawing and painting tools, some of which are illustrated in Figure 8.14, photographic image manipulation tools offer wide scope to the graphic artist and animator. We will look at some of the creative effects obtainable by these techniques.

Paint-system functions

Each of the images in the sequence that follows has undergone a convolution or pixel value-sorting algorithm (ranking) similar to those described in Chapter 4. First, the original image (Figure 8.15).

This first image (Figure 8.16) has been smoothed using a filter similar to the simple box (blur) filter we met in Chapter 4.

Figure 8.14   *Draw and paint functions in 2D graphics*

Figure 8.15   *Original image*

If the maximum value of luminance is taken as *t* and the value of any particular pixel is *a*, Figure 8.17 has been obtained by subtracting *a* from *t* and displaying the new value.

Figure 8.18 is generated by repeating the pixel value in the top left hand of each mosaic block and repeating it over eight horizontal and vertical pixel addresses. It looks a little like a mosaic, but the edges aren't distinct enough. If this picture is subjected to an edge enhancement algorithm (as described in Chapter 4), it looks like Figure 8.19.

A visually striking effect (Figure 8.20) is obtained by applying the diffusion filter, which sorts pixel values into blocks of pixels centred on each pixel in turn. Then the colour of the central pixel is replaced randomly from one or other of the values within the block. It generates an effect as if the colours have locally 'run into one another', like an

Figure 8.16   *Image of Figure 8.15 subjected to blur filter*

Figure 8.17   *Image of Figure 8.15 subjected to negative manipulation*

impressionist painting. It is especially effective in colour, in which case the algorithm has to be implemented on each RGB value in turn. (In other words, this is a TrueColor manipulation routine.)

Another visually striking effect is the emboss filter (Figure 8.21), which was explained in Chapter 4. On the simple image illustrated there, it didn't look a very promising manipulation. However, on a photographic image

Figure 8.18   *Image of Figure 8.15 subjected to sample and hold*

Figure 8.19   *Image of Figure 8.18 subjected to edge-sharpening filter*

the effect is as if the image has been embossed into a metal sheet – hence the name.

Each of the image manipulation techniques shown is described as a window function (as described in Chapter 4). In a window function, a block of neighbouring pixels in the original image all relate to the result of the final pixel in the filtered image. These functions are a progression from point functions (like contrast and brightness adjustments), which act on indi-

Figure 8.20   *Image of Figure 8.15 subjected to diffusion routine*

Figure 8.21   *Image of Figure 8.15 subjected to emboss filter*

vidual pixels. A third class of image manipulation functions take as their starting point individual pixels, but instead of manipulating colour values they manipulate pixel positions. These are termed spatial transformations.

Three important types of spatial image transformation are translation, rotation and zoom. The maths of each of these transformations is illustrated in Figure 8.22. In each case the transformation is assumed about the origin; if this is not the case, it is a relatively easy matter to

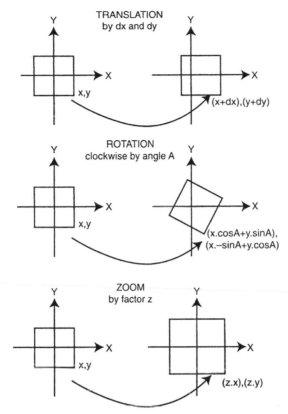

Figure 8.22   *Mathematics of spatial image translations*

include a false origin in the calculations. A complication arises in spatial transformation in bitmap images due to the precise position of calculated points not coinciding with a discrete location within the bitmap raster. One option is to 'round-up' or 'round-down' to the nearest available pixel location, but this can create severe image distortion, especially on small images. A superior (but still far from perfect) technique embodies interpolating the precise position by averaging the transformed pixel value with the value of the pixel at the nearest discrete pixel location and displaying the result. What is really required is a technique that 'up-samples' the image spatially, performs the transformation and then applies a spatial filtering algorithm to 'down-sample' to the required pixelization.

Commercial paint programs invariably permit areas of the image to be selected and cut away to a clipboard for subsequent spatial transformation by translation, zoom and rotation. Figure 8.23 illustrates how paint functions may be used in 2D animation: At the top of the frame is a background. Sometimes backgrounds are drawn by dedicated background

Figure 8.23   *Paint functions in 2D animation*

artists, but in this case I used CorelDRAW to create the illustration. The car was scanned in from a photograph and cut out using the flexible editing tool. By scaling and translating the image several times, a number of key frames were produced to demonstrate how the scanned image may be overlaid on top of the background. Remember that for good movement quality it would have been necessary to create a series of frames where the images were very much closer together than shown: However, the computer can be made to generate these 'tweens', which are the images in beTWEEN the key frames, in a manner very like that for morphing as described below. The act of overlaying one image on top of the other is known as compositing, and this is covered in the next section.

Compositing

We've already seen, in connection with video, how two images may contribute to the output picture at the same time. In television terms this involves a video special effect called a key. Essentially, keying is a process in which a negative hole is cut electronically in the original picture and video from a picture source or caption source is used to fill the 'key-hole'. The computing industry tends to talk less about keying and more about compositing or matting, but the process is exactly the same.

Morphing and warping

The image manipulation technique known as morphing and its brother warping involve both spatial and pixel colour value transformation. Figure 8.24 illustrates the difference (and similarity) between transition morphing and video dissolve; each transition is shown halfway through (a '50 per cent tween'). It can be seen quite clearly that the dissolve is generated by averaging each pixel value in the start image with the value of its corresponding pixel position in the final image. Each pixel maps onto its analogous pixel. In a transition morph, corresponding points within the

Figure 8.24   *A morph contains spatial and pixel colour value transformation. The dissolve (shown below) contains similar pixel colour transformation but no spatial translation*

start and end image need to be defined (by the morph artist) on the basis of contextual loci within the image. (The success of a morph depends largely how carefully this process has been undertaken.) In Figure 8.24 certain points are highlighted and the vectors joining contextually significant locations in the first image are shown, defining the spatial translations that the morph program has performed on the intermediate image. Of course, it would be theoretically possible for the morph artist to define the translation vector for every pixel within the start and end image. This would result in a highly predictable and excellent morph; however, the process would be unbearably labour-intensive. Instead, the morph program generates a vector 'field' on the basis of the number of key points defined by the artist (although the general rule follows that the more key points the better). Only those pixels that are defined as key points have precisely defined translation vectors. Neighbouring pixels are apportioned vectors based on their proximity to the key points. Pixels at loci well away from key points within the start and end image are simply mixed (as in a dissolve).

A warp is similar to the translation morph, except that the start and end images are the same. Key points are used to define translations of certain image points to new locations in the target image. Exactly the same approach is used to apportion vectors to pixels adjacent to key points. Figure 8.25 illustrates a warp.

Figure 8.25   *A warp effect*

Rotorscoping

It goes without saying that the inspiration for animated movement comes from real life. When Walt Disney's studios were animating *Bambi*, they imported deer and had them graze in the fields outside the animation studio so that the artists could observe them as often as they wished. Sadly, not all animation projects have a budget that will stand that kind of attention to detail! A cheaper technique for capturing natural movement is known as rotorscoping, where the animator uses film (or video) footage of real animal or human movement and, working frame by frame, traces the outline and salient movement features of the real image and uses these as primitives on which to base the final animation.

3D graphics and animation

In 3D graphics and animation we want to be able to view objects created with the system from any position and any orientation – in 3D space, like real objects in the real world. For this reason, the first step in a 3D graphics system is a methodology for describing the objects. This description takes the form of a numerical model of both the environment and the objects. This model is termed a 3D artificial environment, or 3Da.e. for short. Each object in the 3Da.e. has height, width and depth. These are represented as dimensions on a three-space Cartesian co-ordinate system. Horizontal dimensions are in terms of x, vertical in terms of y and depth in terms of z. The origin of the co-ordinate system is defined by the triplet (0, 0, 0). It is possible to define a unique position in space with respect to this origin by another triplet (x, y, z). In this environment each object is created form a number of polygons; each polygon has a defined number of vertices and each vertex has three co-ordinates (a triplet) specifying a unique position in space. Notice that every object (even curved ones) is constructed from polygons.

For various practical reasons most 3Da.e's actually employ two co-ordinate systems, one related to the 3D world modelled within the system as already described. Co-ordinates expressed in terms of this system are termed world co-ordinates. The second set is an 'observer-centric' system (known as camera co-ordinates). This apparent complication arises because the aim of the 3D graphics and animation program is to create and manipulate images; it is therefore sensible to operate in terms of the world viewed by an imaginary camera at the observer's position – and hence to use a camera-based co-ordinate system. However, not all descriptions of the 3Da.e. will be particularly convenient in terms of the camera position, so in these cases a world co-ordinate system is employed. It was noted, in relation to spatial image transformations, that false origins could be employed to change (for instance) the centre of rotation. Essentially, the same process is employed in a 3D system to effect translation between one origin and the other. (This may involve both 3D translation of the co-ordinate system centre and rotation of the axis.) As we saw earlier, even planar translation and rotation involve addition and multiplication; not surprisingly, 3D manipulations involve the same techniques but require a greater number of calculations to be performed. Written out 'long hand' these calculations would look very confusing, so the mathematical convention of matrices is used to define the spatial transformations in a 3D system.

Matrices

A matrix is simply an array of numbers arranged in rows and columns to form a rectangular array; a matrix having *m* rows and *n* columns is known as an '*m* by *n* matrix'. There is no arithmetical connection between the elements of a matrix, so it's impossible to calculate a 'value' for it. When referring to a matrix (and to avoid writing it out every time in full) it's possible to denote it by a single letter enclosed in square brackets. So a matrix can be written out as

$$\begin{matrix} a & b & c \\ d & e & f \\ g & h & i \end{matrix}$$

and referred to as [*a*].

Matrices (plural of matrix) may be scaled, added or multiplied together. In each case the following rules apply, taking two generic matrices as starting points; matrix [*p*], which looks like this:

$$\begin{matrix} p & q \\ r & s \end{matrix}$$

and matrix [*t*], which looks like this:

$$\begin{matrix} t & u \\ v & w \end{matrix}$$

Matrices [*p*] and [*t*] are added together using the following rule:

$$\begin{matrix} p+t & q+u \\ v+r & s+w \end{matrix}$$

The same rule applies for subtraction.

Matrices can be multiplied by a constant (this is also termed scaled, or subjected to scalar multiplication). When a matrix is scaled, each element within the matrix is multiplied by the scaling factor. So *n*[*p*] equals

$$\begin{matrix} np & nq \\ nr & ns \end{matrix}$$

Two matrices can only be multiplied together when the number of columns in the first is equal to the number of rows in the second. They are multiplied together using this rule.

$$[p] \times [t] =$$

$$pt + qv \qquad pu + qw$$

$$rt + sv \qquad ru + sw$$

Notice that each element of the top row of [*p*] is multiplied by the corresponding element in the first column of [*t*], and the products added. Similarly, the second row of the product is found by multiplying each element in the second row of [*p*] by the corresponding element in the first column of [*t*].

These matrix calculation techniques are used extensively in 3D graphics to effect translation, rotation and so on. In each case one matrix is the triplet [*x y z*], which defines a particular co-ordinate.

Translation is performed by matrix addition like this:

$$[x\ y\ z] + [dx\ dy\ dz] = [x + dx,\ y + dy,\ z + dz]$$

Rotation is performed by matrix multiplication by three generic rotation matrices. Rotation about the z axis (sometimes called roll) is achieved by multiplication of the triplet by the (3 by 3) matrix:

$$\cos(A) \qquad \sin(A) \qquad 0$$

$$-\sin(A) \qquad \cos(A) \qquad 0$$

$$0 \qquad\qquad 0 \qquad\qquad 1$$

(The similarity with the rotation calculations above is obvious, but notice that the rotation here is anti-clockwise and is positive – in line with mathematical convention.) Rotation about the *y* axis (yaw) is performed by multiplying each co-ordinate triplet by:

$$\cos(A) \quad 0 \quad \sin(A)$$

$$0 \qquad\quad 1 \quad 0$$

$$\sin(A) \quad 0 \quad \cos(A)$$

and rotation about the *x* axis (pitch) by multiplying by:

$$1 \quad 0 \qquad\qquad 0$$

$$0 \quad \cos(A) \qquad \sin(A)$$

$$0 \quad -\sin(A) \qquad \cos(A)$$

Increasing the overall volume of a body is achieved by multiplying each triplet with a scaling factor. In each case, adoption of matrix notation ensures that all the correct multiplications have been performed and calculated in the appropriate order.

Imaging

We have specified a co-ordinate system that defines a 3D environment, a hierarchical structure so that we can build objects (bodies) within the environment based on polygons, in turn specified by vertices, in turn specified by triplets. Take a simple example; a cube. A cubic body would be defined in terms of six polygons; the position of each would be determined by four vertices; each vertex would be defined by a triplet. Furthermore, with the matrix transformations described above we are in a position to manipulate the position and orientations of bodies within the environment. The irony is, having once created a mathematical three-dimensional world, the 3D graphics system (because it has to display images on a 2D television screen) has to turn everything back to two dimensions again! Therefore, the final stage of the 3Da.e. manipulation involves finding screen co-ordinates of each point; given its co-ordinates in 3D space. Several alternative transformations are available that produce a two-dimensional image. The one chosen in most graphics systems is called a 'perspective transformation'. In a perspective transformation, a point in space is designated as a focal point and a plane is introduced between the focal point and the scene to be viewed. For every point on the 3Da.e. (or portion of it to be viewed) a straight line is projected from that point to the focal point. The position on the plane where the line joining the point and the focal point penetrates is the projection of the point in the 3Da.e. If this process is carried out for every point in the scene, and the points on the screen are joined in the same order they are joined in the scene, then a 2D perspective image is produced of the 3D scene. Practically, perspective transformation turns out to be simply another transformation; this time about the camera-centric co-ordinate system.

For a co-ordinate triplet (expressed in terms of the camera co-ordinate system)

$$[xC \; yC \; zC]$$

the projection of this point (i, j) on the screen (at the origin of the camera co-ordinate system) is given by:

$$i = xC / \{1 + (zC/D)\}$$

and

$$j = yC / \{1 + (zC/D)\}$$

where D is the distance of the focal point behind the screen plane (see Figure 8.26).

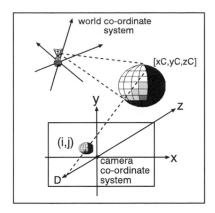

Figure 8.26 *A 'curved' object modelled in our 3Da.e. which is illum-inated by a source of light and each polygonal face is reflecting light back to the focal point of the viewing system. The intercept in each ray with the plane of the screen is a pixel within our final image*

Light

Unfortunately, as it is, the world we have created is of entirely academic interest because we are unable to see any of the polygonal bodies we might wish to furnish it with, lacking, as we do, any form of illumination. Most practical artificial light sources are so-called point sources. The light from a point source is assumed to emerge from an infinitesimally small region of space and spread out evenly in all directions in straight-line rays. As a result, the further the light spreads out from its original point of origin before it reaches an object, the less light falls upon a surface of the object. A standard incandescent light bulb can be thought of as a point source. Imagine fitting a 100 W light bulb in the centre pendant light of a small, whitewashed room. The room would appear fairly bright, because the walls would reflect the relatively large amount of light energy falling upon each square area of their surface. Now imagine fitting the same bulb in an enormous whitewashed room; the space would appear very dim, not because the walls weren't reflecting the light as efficiently, but because the amount of light emitted from the bulb would be diluted or spread out over a much greater area of wall.

The solar system makes the dimensions of even the largest room look positively atomic, so to all intents and purposes illumination by the sun doesn't appear to follow the same pattern. The light from the sun is assumed to reach the earth by parallel rays. Consequently, the degree of illumination doesn't change depending on the distance. (Other sources of parallel rays exist; for instance, a laser and a theatre spotlight.) The amount of light energy (or illumination, E) falling on a surface of an object lit by a source of parallel light rays is a function solely of the luminous intensity of the light source (I) and the angle the surface makes to the parallel rays (i), so:

$$E = I \times \cos(A)$$

In the case of an object lit by a point source, the equation has to take account of the dissipation of the available energy over a given surface area depending on its distance from the light source. In this case the illumination is given by:

$$E = I \times \cos(A)/r^2$$

where r is the distance (radius) of the object from the light source. The illumination can be seen to fall off with the square of this distance.

Our 3D world is now equipped with light (or lights) which obeys physical rules, but we still shouldn't be able to see anything in it because each of the polygonal surfaces has yet to be attributed a luminance. The luminance of a surface is defined as the luminous energy coming from a surface; in other words, the amount of light a particular surface reflects. Different materials reflect differently. Luminance should be distinguished from illumination, which refers to the amount of light incident on the surface. Difference in luminance is due to difference in reflectance factor, R. For objects which are opaque (i.e. non-shiny), the luminance (Y) of the so-called diffuse light scattered from the surface is given by:

$$Y = R \times E$$

When the surface is shiny, different rules apply. Consider the shiniest surface of all, a mirror. In the case of a perfect mirror; if a ray of incident light strikes the mirror at an angle a to the surface of the mirror, it 'bounces off' and leaves the mirror (is reflected) at an identical angle to the plane of the mirror surface, as illustrated in Figure 8.27.

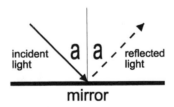

Figure 8.27   *Reflection at a shiny surface*

Due to imperfections in the surface of the mirror and other physical effects, the situation illustrated in Figure 8.27 isn't quite so straightforward. In fact, the incident light is reflected over a number of angles. This effect is known as diffuse reflection. As you may have guessed, there exists a continuum of reflection from the perfect (regular) to the entirely diffuse, as in the case of an opaque surface.

Figure 8.28 illustrates a curved object modelled in our 3Da.e. which is illuminated by a source of light, and each polygon is reflecting this light back to the focal point of the viewing system. The intercept in each ray with the plane of the screen is a pixel within our final image. The polygonal nature of the supposedly curved surface is pretty obvious. Several different mathematical strategies exist for eliminating the faceted nature of polygonal models, and these are termed polygonal shading techniques. Every image point on each polygon is subject to the same lighting, and therefore every pixel within each polygon is the same colour. In the technique known as Gouraud shading, the colours are interpolated to eliminate the distinction between the individual faces.

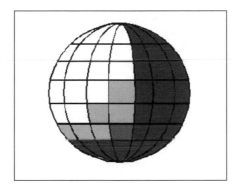

Figure 8.28 *Polygonal 'sphere'*

Ray tracing

Because each object within the scene possesses reflectance, each acts as its own small light source. Once light from the primary source has reflected off a surface it leaves that polygonal face, possibly over a wide range of angles if the surface is opaque. On the other hand, if the surface is shiny, it may reflect over a precisely defined range of angles. Perhaps the ray of light becomes coloured due to its interaction with the reflecting surface. The process whereby each light ray is traced along its journey through the 3Da.e., and the effect of its reflections and their subsequent role as illuminants of other objects within the environment, is known as ray tracing. The calculation of lighting effects is one of the most computationally intensive parts of any 3D graphics system. The mathematical calculations for this and other aspects of the final image production (like Gouraud shading) take place in a process known as rendering, which may take many hours even with a very powerful computer.

Hard disk technology

The spread of computing technology into all areas of modern life is so obvious as to require no introduction. One consequence of this is the drift towards the recording of entertainment media (audio/video – albeit digitally coded) on computer-style hard disks, either within a computer itself, with the machine's operating system dealing with the disk management, or within bespoke recording hardware utilizing disk technology. But first a question; why disks and not tape?

The computer industry all but stopped using tape technology many years ago. The reason is simple. Whilst a tape is capable of storing vast quantities of data, it does not provide a very easy mechanism for retrieving that data except in the order that it was recorded. The issue is coined in the computing term 'access time'. To locate a piece of data somewhere on a tape may take several minutes, even if the tape is wound at high speed from the present position to the desired location. This is really not such an issue for entertainment, since a television programme is usually intended to be watched in the order in which it was recorded. However, it is an issue for vision editors and producers, because during an edit the tape may have to be rewound hundreds or even thousands of times, thereby reducing productivity as well as stifling the creative process. Far better, then, to enjoy the benefits of computer disks which, because the data is all available at any time spread out as it were 'on a plate' (quite literally – see Figure 8.29), make all the recorded signals available virtually instantaneously.

Figure 8.29   *Hard disk-drive construction*

Winchester hard disk drive technology

Think of disk drive technology as a mixture of tape and disk technology. In many ways, it combines the advantages of both in a reliable, cheap package. In a disk drive, data is written in a series of circular tracks, a bit like a CD or an analogue LP, but not as a wiggly track (as in the case of the LP) or as a series of physical bumps (as in the case of the CD); rather as a series of magnetic patterns. As in the case of the CD and the record, this implies that the record and replay head must be on a form of arm that is able to move across the disk's surface. It also implies (and this it has in common with the CD) the presence of a servo-control system to keep the record/replay head assembly accurately tracing the data patterns, as well as a disk operating system to ensure an initial pattern track is written on the disk prior to use (a process known as formatting). Like an LP record, in the magnetic disk the data is written on both sides. The process of formatting a disk records a data pattern onto a new disk so that the heads are able to track this pattern for the purposes of recording new data. The most basic part of this process is breaking the disk into a series of concentric circular tracks. Note that in a disk drive the data is not in the form of a spiral, as it is on a CD. These concentric circular tracks are known as tracks, and are sub-divided into sections known as sectors.

In a hard drive, the magnetic medium is rigid and is known as a platter. Several platters are stacked together, all rotating on a common spindle, along with their associated head assemblies, which also move in tandem (see Figure 8.29). Conceptually, the process of reading and writing to the disk by means of a moveable record/replay head is similar to that of the more familiar floppy disk. However, there are a number of important differences. Materially, a hard disk is manufactured to far tighter tolerances than the floppy disk, and rotates some 10 times faster. Also, the head assembly does not physically touch the disk medium but instead floats on a microscopic cushion of air. If specks of dust or cigarette smoke were allowed to come between the head and the disk, data would be lost, the effect being known as a head crash. To prevent this, hard-drives are manufactured as hermetically sealed units.

Other disk technologies

Read/write compact disk (CD-R) drives are now widely available and very cheap. CD-R drives are usually SCSI based, so PCs usually have to have an extra expansion card fitted to provide this interface (see below). Recordable CDs rely on a laser-based system to 'burn' data into a thermally sensitive dye which is coated on the metal medium. Depending on the disk type, once written the data cannot be erased; other CD-RW disks are read and write (RW), and may be erased and used again. Software exists (and usually comes bundled with the drive) which enables the drive to be used as a data medium or an audio carrier (or sometimes as both). There are a number of different variations of the standard ISO-9600 CD-ROM. The two most important are the (HFS/ISO) hybrid disk, which provides support for CD-ROM on Mac and PC using separate partitions, and the mixed mode disk, which allows one track of either HFS (Mac) or ISO-9600 information and subsequent tracks of audio.

A number of alternative removable media are available and suitable for digital audio and video use; some based on magnetic storage (like a floppy disk or a Winchester hard-drive) and some on magneto-optical techniques – nearer to CD technology. Bernoulli cartridges are based on floppy disk, magnetic storage technology. Access times are fast enough for compressed video and audio applications; around 20 ms. SyQuest is similar; modern SyQuest cartridges and drives are now available in up to several Gbytes capacity and 11 ms access times, making SyQuest the nearest thing to a portable hard-drive. Magneto-optical drives use similar technology to CD; they are written and read using a laser (Sony is a major manufacturer of optical drives). Sizes up to several Gbytes are available, with access times between 20 and 30 ms.

Hard drive interface standards

There are several interface standards for passing data between a hard disk and a computer. The most common are: the SCSI or small computer system interface, the standard interface for Apple Macs; the IDE or integrated drive electronics interface, which is not as fast as SCSI; and the enhanced IDE interface, which is a new version of the IDE interface that supports data transfer rates comparable to SCSI.

IDE drives

The integrated drive electronics interface was designed for mass storage devices in which the controller is integrated into the disk or CD-ROM drive. It is therefore a lower cost alternative to SCSI interfaces, in which the interface handling is separate from the drive electronics. The original IDE interface supports data transfer rates of about 3.3 Mbytes per second and has a limit of 538 Mbytes per device. However, a newer version of IDE, called enhanced IDE (EIDE) or fast IDE, supports data transfer rates of about 12 Mbytes per second and storage devices of up to 8.4 Gbytes. These numbers are comparable to what SCSI offers. However, because the interface handling is handled by the disk-drive, IDE is a very simple interface and does not exist as an inter-equipment standard; that is, you cannot connect an external drive using IDE. Due to demands for easily upgradable storage capacity and for connection with external devices such as recordable CD players, SCSI has become the preferred bus standard in audio/video applications.

SCSI

An abbreviation of small computer system interface and pronounced 'scuzzy', SCSI is a parallel interface standard used by Apple Macintosh computers (and some PCs) for attaching peripheral devices to computers. All Apple Macintosh computers starting with the Macintosh Plus come with a SCSI port for attaching devices such as disk drives and printers. SCSI interfaces provide for fast data transmission rates; up to 40 Mbytes per second. In addition SCSI is a multi-drop interface, which means you can attach many devices to a single SCSI port.

Although SCSI is an ANSI standard, unfortunately, due to ever-higher demands on throughput, SCSI comes in a variety of 'flavours'! The following varieties of SCSI are currently implemented:

- SCSI-1: uses an 8-bit bus, and supports data rates of 4 Mbytes/s.
- SCSI-2: same as SCSI-1, but uses a 50-pin connector instead of a 25-pin connector. This is what most people mean when they refer to plain SCSI.
- Fast SCSI: uses an 8-bit bus, and supports data rates of 10 Mbytes/s.
- Ultra SCSI: uses an 8-bit bus, and supports data rates of 20 Mbytes/s.
- Fast wide SCSI: uses a 16-bit bus and supports data rates of 20 Mbytes/s.
- Ultra-wide SCSI: uses a 16-bit bus and supports data rates of 40 Mbytes/s; this is also called an SCSI 3.

Fibre channel

Fibre channel is a data transfer architecture developed by a consortium of computer and mass storage device manufacturers. The most prominent fibre channel standard is fibre channel arbitrated loop (FC-AL), which was designed for new mass storage devices and other peripheral devices that require very high bandwidth. Using an optical fibre to connect devices, FC-AL supports full-duplex data transfer rates of 100 Mbit/s. With this sort of data-rate, it's no surprise that fibre channel has found its way into the modern studio; so much so that FC-AL is expected eventually to replace SCSI for high-performance storage systems.

Firewire

The Firewire (IEEE 1394 interface) is an international standard, low-cost digital interface that is intended to integrate entertainment, communication and computing electronics into consumer multimedia. Originated by Apple Computer as a desktop LAN, Firewire has been developed by the IEEE 1394 working group. Firewire supports 63 devices on a single bus (SCSI supports 7, SCSI Wide supports 15), and allows busses to be bridged (joined together) to give a theoretical maximum of thousands of devices. It uses a thin, easy to handle cable that can stretch further between devices than SCSI, which only supports a maximum 'chain' length of 7 meters (20 feet). Firewire supports 64-bit addressing with automatic address selection and has been designed from the ground up as a 'plug and play' interface. Firewire originally only handled 10 Mbytes per second, but has a long term bandwidth potential of over 100 Mbytes/s. Much like LANs and WANs, IEEE 1394 is defined by the high level application interfaces that use it, not a single physical implementation. Therefore, as new silicon technologies allow high higher speeds, longer distances, IEEE 1394 will scale to enable new applications. (IEEE 1394 is discussed as a consumer digital interconnect in Chapter 11.)

RAID

RAID is an acronym for redundant array of independent (or inexpensive) drives. RAID technology concerns the use of storing and retrieving data on an array of multiple hard disks as opposed to a single hard drive. An array of RAID disks always has a controller built-in; the computer just 'thinks' it's talking to a normal disk drive. All the 'clever bit' goes on inside the RAID array control device. Why use RAID? There are two reasons. The first is speed. Multiple disks, accessed in parallel, give greater data throughput (write/read speed) than a single disk. The second reason is reliability. With a single hard disk, you cannot protect yourself against catastrophic disk failure. Anyone who has experienced a total disk-drive crash will know the agony of installing a new drive; re-installing the operating system and restoring files from backup tapes (assuming you've been careful enough to make these!) With an appropriate RAID disk array, your system can stay up and running when a disk fails. Moreover, RAID controllers are intelligent enough that, should one disk fail and be replaced with a virgin drive, it will rebuild the original array.

To some extent these two objectives are contradictory, and the term RAID covers several different arrangements, each with a different emphasis on speed versus reliability. Originally RAID came in five different varieties, termed RAID 1 to RAID 5. Some proved more useful than others. Recently RAID definitions have been extended (corrupted?), so you will sometimes see references to RAID 0, which is a scheme that uses multiple disks with no redundancy (and is therefore actually not RAID at all, but AID!). Similarly, you may see references to RAID 35, which is a mixture of RAID 3 and 5. Each of the five original RAID schemes is described below.

RAID 1 (mirroring)

RAID 1 is usually called 'mirroring', and its emphasis is on data security. All disks in the array are arranged in pairs, and RAID 1 provides complete redundancy by writing identical copies of all data on these pairs of disks. For all its 'belt and braces' approach, RAID 1 still offers some increase in speed because writing to the disks can be done in parallel, whereas reads can be interleaved.

RAID 2 (bit striping with error correction)

Unlike parallel RAID 1, RAID 2 works in series. The controller writes sequential blocks of data across multiple disks. Each sequential block is termed a stripe, and the size of the block is termed the stripe width. In RAID 2, the stripe width is 1 bit only. A RAID 2 system would therefore have as many data disks as the word size of the computer, and every disk access must involve every disk. In addition, RAID 2 requires the use of extra disks to store error-correction codes for redundancy. With 32 data disks, and a few parity disks thrown-in for good measure, it's not surprising that RAID 2 has never been considered a practical option.

RAID 3 (bit striping with parity)

RAID 3 is very similar to RAID 2, except that only one extra disk is used to store simple parity data. This parity disk is written with data derived quickly and simply from the 8, 16 or 32 data bits on the other drives. This only works because the disk controller of the drive which experiences the missing bit is able to report that it has had a data read error. Knowing which disk's data is missing, the RAID controller can reconstruct the original data. For instance, imagine we write the byte 10010001 to eight RAID 3 drives. Assuming we use a simple even-parity scheme, we would write 0 as the data on the ninth parity drive because there is an odd number of ones in the original byte. So we would actually write the following across all nine drives:

$$1 \quad 0 \quad 0 \quad 1 \quad 0 \quad 0 \quad 0 \quad 1 \quad (0)$$

where the (0) is the value on the parity drive.

Now, suppose in a subsequent read command we receive the following:

$$1 \quad 0 \quad 0 \quad * \quad 0 \quad 0 \quad 0 \quad 1 \quad (0)$$

We have an even number of ones, but, because parity is 0 (or NOT EVEN), we know that the failed bit must have been a one.

The one drawback of RAID 3 is that it must read all data disks for every read operation. This works best on a single-tasking system with large sequential data requirements – for example, a broadcast quality video-editing system, where huge video files must be read sequentially.

RAID 4 (striping with fixed parity)

RAID 4 is the same as RAID 3, except that the stripe widths are much greater; the intention being that individual read requests can be fulfilled from a single disk. However, this isn't the case, because each read and write request has to access the single parity disk. This is such a drawback that RAID 4 is never implemented.

RAID 5 (striping with striped parity)

RAID 5 uses large stripe widths and also stripes the parity across all disks. This scheme provides all the advantages of RAID 4, and it avoids the bottleneck of a single parity disk.

Media server

Take a broadcast news studio. A few years ago these studios were a hive of frenetic activity; operators dashed backwards and forwards with tape cassettes, rushing to an edit, back from an edit, running to playout. Journalists rushed backwards and forwards with floppy disks with bits of stories on. Graphics people bolted about with logo designs, maps, titles and captions on floppy disks too – but usually not the same sort of floppy disk used by the journalists. Everyone seemed to be in motion, clutching little 'packages' of information. You could think of such a system as a high bandwidth 'sneaker-net'! Now imagine if all that information, not just the text but the pictures, the sound, the logos and graphics, the schedules, the time-sheets, everything, was held in one central store. What's more, imagine if everybody had access to that store – so a journalist could check a map out, a graphics operator could check a spelling. And the editor could sit, like a queen bee, watching, monitoring and selecting the best input and slotting it into an automatic playout sequence. That's the aim of a media server. As you'd expect, a media server is tailored towards two vital performance criteria; massive data storage capability and excellent connectivity. Normally a media server can accept several full resolution video channels simultaneously, or literally hundreds of compressed video channels. It must also support multiple networking protocols, including Ethernet and ATM, as well HiPPI.

Open media framework

As in so many areas where computer technology is changing the way work gets done, it's not long before the professionals who use the equipment start to yearn for standards so that they can transfer data between different applications and platforms. Now that video post-production is increasingly executed on computer-based editing systems, users are expressing just this need. That's the idea behind the open media framework (OMF) interchange. Essentially, OMF defines a cross-platform file format that allows editors, animators and other television and film professionals to create (for instance) a project on a Mac and move it over to a Windows PC; or permit an artist to create graphics on an SGI and move it across to perform compositing on a Mac. Of course, file format compatibility is but one issue here; even if file formats are compatible, there's the problem of physically moving data across platforms and problems of different types to be moved over different networks.

Virtual sets

A powerful use of virtual reality-type technology, applied to broadcast television, exists in the technique known as virtual sets. This method is, essentially, an extension of chroma-key, in which a subject is shot in front of a coloured flat or cyclorama and, using a colour-separation technique, their image is superimposed on another television picture, thereby giving the impression that the presenter is somewhere other than in a studio. Virtual sets take this technique several steps further by attaching movement sensors to the televising camera and applying the data so obtained to a high speed computer, programmed to model and render a 3D artificial environment (3Da.e.) at very high speed. In this manner the presenter may move about a virtual environment and the camera-person may zoom-in, track, or crab, and the appropriate spatial transformations will occur both to the image of the presenter (due to the real-world spatial relationships involved) and to the virtual environment, or virtual television set (due entirely to high-speed computations performed in transforming and projecting the 3Da.e.). The potential cost savings from such a technology are pretty obvious, especially in television drama productions.

The master control room

When television was mostly live, each individual studio control room fed a master control room (MCR), whose main function was to select between the output of the individual studios. Over and above this function, the MCR often added a degree 'continuity', comprising a static card comprising the station logo and a disembodied voice known as a 'voice over', announcing the next programme and perhaps the rest of the evening's viewing. With the advent of the transmission of largely recorded material, the MCR became the obvious place to house the video tape, automated video cassette machines and later, video servers, which provide the majority of the television output. But the traditional role of the MCR remains in providing continuity between each programme segment whether this be by video transition, voice over, or by the modern replacement of the static station card, a station 'generic' – a short piece of video designed to brand the television output. The equipment designed to accomplish these master control and channel branding tasks is known as the master control vision mixer and an example of a product (the Presmaster from Miranda Technologies Inc.) is shown in Figure 8.30. The Presmaster is designed to fulfil all the tasks of master control and channel branding required by a modern television station including the selection of individual programme outputs, either manually or under automation, the addition of voice-overs, of station generics and the addition of a permanent or semi-permanent video key restricted to one or another corner of the television screen known as the station logo.

Figure 8.30

The Presmaster is particularly interesting because it may be used to control multiple station outputs from one control panel by means of its distributed architecture of its video and audio processing electronics illustrated in Figure 8.31. This is a necessary requirement to avoid spiralling operator costs in the multi-channel digital world.

Automation

Once again, in the interests of controlling operating costs and in providing slick master control and channel branding, most of these

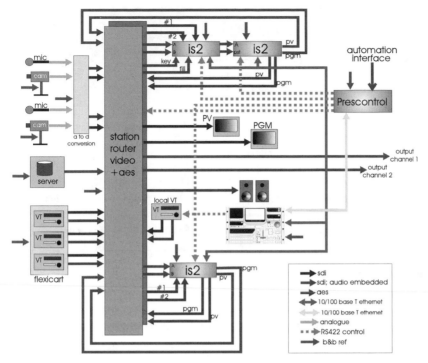

Figure 8.31

operations have become automated, the video cassette machines, the vision mixer functions and the video servers being controlled from a sophisticated time-switching device known as a television automation system. The automation system software may run on dedicated micro-processor hardware or on a non-dedicated microcomputer platform: its role is to send control signals (usually via RS422 connections) to the various pieces of equipment housed in the MCR under the auspices of the channel automation 'schedule'. This schedule is authored each day and is very often automatically derived from the television traffic system – the management system for the planning of television programming and commercial scheduling derived in turn from advertisement of sales. An example of an on-screen schedule is given in Figure 8.32. You can clearly see the individual programme segments as well as automated control functions relating to the type of transition between each programme segment as well as the addition of voice overs and keys. The schematic of simple single channel television station controlled by automation is given in Figure 8.31. Notice that all the signal processing is still performed in the SDI/AES domain with the MPEG encoder at the end of signal chain.

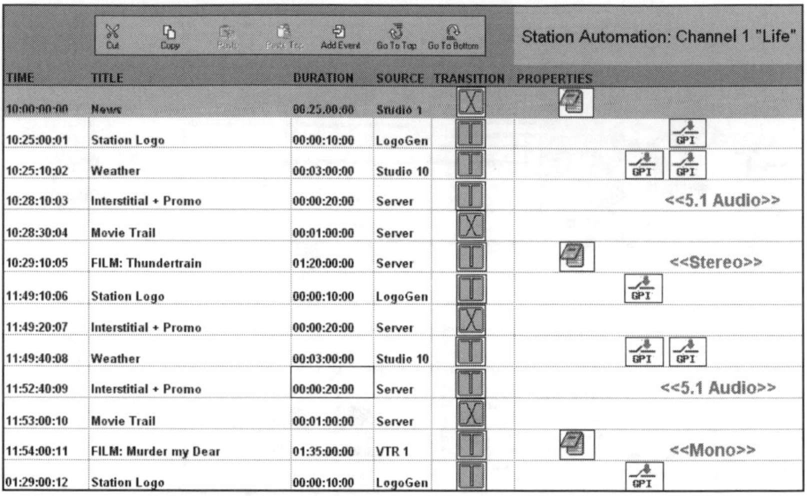

TIME	TITLE	DURATION	SOURCE	TRANSITION	PROPERTIES
					Station Automation: Channel 1 "Life"
10:00:00:00	News	00:25:00:00	Studio 1	[X]	[icon]
10:25:00:01	Station Logo	00:00:10:00	LogoGen	[II]	GPI
10:25:10:02	Weather	00:03:00:00	Studio 10	[II]	GPI GPI
10:28:10:03	Interstitial + Promo	00:00:20:00	Server	[II]	<<5.1 Audio>>
10:28:30:04	Movie Trail	00:01:00:00	Server	[X]	
10:29:10:05	FILM: Thundertrain	01:20:00:00	Server	[II]	[icon] <<Stereo>>
11:49:10:06	Station Logo	00:00:10:00	LogoGen	[II]	GPI
11:49:20:07	Interstitial + Promo	00:00:20:00	Server	[X]	
11:49:40:08	Weather	00:03:00:00	Studio 10	[II]	GPI GPI
11:52:40:09	Interstitial + Promo	00:00:20:00	Server	[II]	<<5.1 Audio>>
11:53:00:10	Movie Trail	00:01:00:00	Server	[X]	
11:54:00:11	FILM: Murder my Dear	01:35:00:00	VTR 1	[II]	[icon] <<Mono>>
01:29:00:12	Station Logo	00:00:10:00	LogoGen	[II]	GPI

Figure 8.32   *Automation schedule*

Editing and switching of MPEG-II bitstreams

Normally MPEG-II compressed video will form an element of a MPEG transport stream. A switch can be performed directly between transport streams on transport packet boundaries; this is referred to as splicing. However, splicing has some severe limitations. First, the splice point must correspond to the end of an I- or P-frame in the 'old' bitstream and the splice point must correspond to the start of an I-frame in the 'new' bitstream. For many MPEG-II applications, this will mean that splicing can only be performed to a resolution of about half a second. Similarly, the buffer of a downstream decoder must be at a particular state at each splice point (including unused ones). This causes rate control restrictions on the coder(s) producing the bitstreams to be spliced, which may lead to loss of quality if a large number of splice points are to be used. Moreover, transitions other than cuts (e.g. cross-fades) are not possible. These restrictions limit the range of applications for splicing of transport streams.

One option (and currently the favourite technique) is to decode, switch and re-code. By this means, the switch points can occur on any frame, and the switching imposes few constraints on the incoming bitstreams. In addition, existing ITU-R Rec.601 equipment such as vision mixers can be used, and other transitions, such as cross-fades, can be used between in addition to simple cuts. This approach also allows switching between different types of compressed signals and between compressed and uncompressed signals. However, this simple approach will lead to loss of picture quality due to cascaded coding, particularly when the recoder uses a different GOP phasing to the original coder. But even if means can be taken to prevent this happening, degradation will be caused by the use of different motion vectors, coding modes and quantizer setting on recoding.

The ATLANTIC Project

A major part of the pan-European (and EEC funded) ATLANTIC Project was to develop techniques to allow MPEG-II bitstreams to be used throughout the programme chain. The techniques developed make use of an additional output produced by an otherwise standard MPEG decoder known as the 'info-bus'. The info-bus contains information on how the bitstream was coded, e.g. picture type, prediction mode, motion vectors. This information is 'passed around' the otherwise standard 601 switching equipment and is used by a co-operating, downstream coder when recompressing the signal. Because the info-bus contains the vectors, the recoder does not need a full motion estimator. It can be shown that if all coding parameters and decisions are kept the same, additional generations of MPEG video coding can be performed transparently.

'Mole'

In a later development, the info-bus from the MPEG-II decoder is converted to a signal known as a mole which is 'buried' into the SDI signal. This signal is in a form that enables it to be multiplexed invisibly into the decoded SDI video signal, and is switched transparently along with the SDI in the vision mixer. The mole is then converted back to an info-bus prior to being used for recoding the switched video.

9
The MPEG multiplex

A 'packetized' interface

We shall see that the MPEG system for transmission of digital television relies heavily on the idea of small 'packets' of information moving together in a much larger data 'pipe'. This idea originated in telecommunications, where hundreds – perhaps thousands – of telephone calls travel together down one coaxial cable or optical fibre. We have seen something similar to this in relation to MADI, in Chapter 3, although in that case each multiplexed channel was ascribed a particular time position in the multiplex. A packetized system is much cleverer. Consider the metaphor of packets for a moment. Think of the London to Glasgow post train. The train may carry thousands of parcels (packets) all jumbled together; toy boats, books, jewels – each one destined for a different place. How do they get 'un-jumbled'? Because each one has a label clearly stating the address to which it should be delivered. In just the same way, in a packetized interface, each packet has to contain information allowing it (and only it) to be selected from the melee of other information. This electronic 'address label' is known as a header and, just as the postal system likes to have post codes and insist that a parcel is clearly labelled, the definition and structure of header information is a very important part of the definition of a packetized interface. Packet headers may also contain information concerning how the particular packet should be treated, in the same way a real parcel might bear the stickers 'This way up' or 'Fragile'.

Now, you can't just stick a label and some stamps on a child's tricycle and expect the Post Office to deliver it! You have to wrap it – you have to deliver it in a form that the post office can accept. The same principle applies to a packetized interface. Packets have defined sizes and forms, and each may be 'wrapped' with a data protection scheme (parity bits – or electronic bubble-wrap!) so that they arrive in one piece at their destination. There is, however, one important difference with electronic packets. In a packetized interface, often a much larger element is broken down into small parcels for transportation; the tricycle is broken into bits, each one wrapped and addressed and sent as a separate packet. Unlike a real tricycle, in electronic form we can easily glue the original information stream together from all the small packets that arrive!

This then is the essence of a digital television multiplex; a single data stream (bitstream) which is composed of packets of morsels of one or more television (and radio) programmes, each of a prescribed form, with prescribed labels. It is a signal of this type that energizes the cable modulator or the satellite transponder or modulates a terrestrial television carrier signal. In this way a single carrier can be used to carry several channels of digital television. A digital service of this type is termed a 'bouquet'.

Deriving the MPEG-II multiplex

Figure 9.1 illustrates the organization of a simple digital television transmission service. The data transport mechanism, the MPEG-II transport stream, is based on the use of fixed-length (188 byte) transport stream packets identified by headers. Each header identifies a particular application bitstream (also called an elementary bitstream, or ES) which represents real, coded, 'entertainment' information. Signal types supported include video, audio, data, programme and system control information, as well as the possibility of other 'private' information carried for various associated services and for descrambling data 'keys'. The elementary bitstreams for video and audio are themselves wrapped in a variable-length packet structure called the packetized elementary stream (PES) before transport processing. Looking at Figure 9.1, note the DTV coding hierarchy: Individual coded PESs with related content (video, associated audio and data) combine together to produce a single program transport multiplex. These, in turn, are combined to form a system transport multiplex. (Note that, although the coders are shown producing program transport multiplex which are subsequently combined to form a system multiplex, practical coders may generate system multiplex directly from multiple elementary streams.)

The PES packet format

PES packets are of variable length, with a maximum size of 65536 bytes. (Note that this is much longer than a MPEG transport packet – in other words, an elementary stream packet will be sub-divided into many transport packets.) A PES packet consists of a header and a subsequent

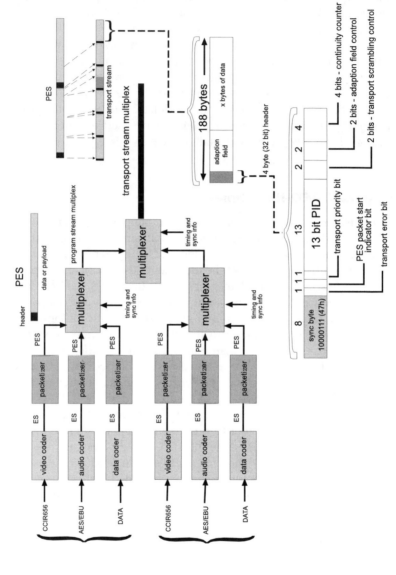

Figure 9.1 *Derivation and structure of the MPEG multiplex*

payload. The payload is created by the application encoder, and is a stream of contiguous bytes of a single elementary stream. The PES header contains various flags and associated data. Of these, PTS and DTS are especially important. The presentation time stamp (PTS) informs the decoder of the intended time of presentation of a presentation unit, and the decoding time stamp (DTS) is the intended time of decoding of an access unit. These flags are used for synchronizing audio, video and other data at the receiver, and must be sent relatively often (at less than 0.7 s intervals); they are discussed in more detail below.

Transport stream

The transport stream consists of 188-byte fixed-length packets with a fixed and a variable component to the header field as illustrated. The content of each packet is identified in the packet header. The packet header structure is made up of two parts; a combination of a 4-byte, fixed-length part and a variable-length adaptation part. Some of the important functions of the header are described here.

Packet synchronization

Packet synchronization is affected by the sync_byte (note the underscore notation, which is standard in the DTV world), which is the first byte in the packet header. The sync_byte has a fixed value (47 h), although this is adapted to two alternating values (47 h and B8 h) later in the coding process for transmission, as we shall see. The sync_byte is used within the decoder to achieve packet synchronization.

Packet identification

After the video and audio signals themselves, the 13-bit header field called the packet identification (PID) field is probably the most important piece of information within the transport packet, providing, as it does, the mechanism for multiplexing and demultiplexing of bitstreams. The PID provides the identification of packets belonging to a particular elementary or control bitstream. (The PID is the equivalent of the postal address label we considered earlier.) The PID field appears in the packet header because, like this, it's always in a fixed place, making the extraction of the packets corresponding to a particular elementary bitstream very simple to achieve once packet synchronization is established in the decoder.

Program association tables and program map tables

How does an MPEG decoder know what PID addresses it ought to be looking for? For this, it uses a series of hierarchical data tables that are transmitted, like video and audio information, in packets. These are used to tell the decoder which television programmes appear on the multiplex and where to find them. There are also other tables that relate to conditional access and to ancillary user data.

The PID contains a very important pointer at PID equal to zero. This field contains the PID address of the program association table, or PAT. The PAT defines various parameters according to each television programme carried on the multiplex, including the program map table (PMT) for each television programme. It is in the PMT that the PID of each elementary stream of each programme is identified. This sounds very complicated, but in fact it's just a string of address pointers such that PID = 0 points to a PID address of the PAT, and the PAT points to PMT, which points to the PIDs of individual elementary streams.

The maximum spacing allowed between occurrences of a program_ map_table containing television programme information is 400 ms. It is important that this information be repeated frequently in order to establish video and audio relatively quickly after a user channel-change command.

Error handling

Error detection at packet level is achieved in the decoder by the use of the continuity_counter field. At the transmitter end, the value in this field cycles from 0 to 15 for all packets with the same PID. At the receiver end, under normal conditions, the reception of packets in a PID stream with a discontinuity in the continuity_counter value indicates that data has been lost in transmission.

The adaptation header

The adaptation_field_ control bits of the transport-level header signals the presence of the adaptation field. This part of the transport header is termed the link header. The adaptation header in the MPEG-II packet is a variable-length field. The use of this field is very varied. It contains further synchronization and timing data, as well as providing a 'stuffing' function in order to load PES packets into the transport stream in a pre-defined way. This level of organization ensures that the transport stream may be manipulated in certain ways in subsequent processing; as we shall see. The adaption header also contains a number of important flags to indicate the presence of the particular extensions to the field.

Synchronization and timing signals

In a DTV system, the amount of data generated for each picture is variable because it is based on a picture coding approach that varies according to a number of picture parameters. For this reason, timing information cannot be derived directly from the start of picture data as it is in analogue television. There is a similar requirement to keep video and its associated audio synchronized. The solution to this is to transmit timing information in the adaptation headers of selected packets to serve as a reference for timing comparison at the decoder. In other words, certain packets are time-stamped and the decoder is responsible for making sure these packets are decoded reasonably close to the required presentation time. But how does the decoder know what time it is? Because it is provided with reference time signals which are included in the multiplex as well. The standard includes two parameters; the system clock reference (SCR) and the program clock reference (PCR).

System and program clock references

An SCR is a snapshot of the encoder system clock that is placed into the system layer of the bitstream. During decoding, these values are used to update the system clock counter within the decoder. The PCR is inserted at the programme layer. (Remember that individual TV signals within the multiplex can have different video references – that is, they are not necessarily 'locked'.) Put simply, the SCR and the PCR are the means by which the decoder knows what time it is and synchronizes its internal 27 MHz PLL. The MPEG-specified 'system clock' runs at 90 kHz. SCR and presentation timestamp values are coded in MPEG bitstreams using 33 bits, which can represent any clock cycle in a 24-hour period.

Presentation timestamps

Presentation timestamps (PTSs) are samples of the encoder system clock that are associated with video or audio presentation units. A presentation unit is a decoded video picture or a decoded audio time sequence. The PTS represents the time at which the video picture is to be displayed, or the starting playback time for the audio time sequence. The PTS is the means by which the decoder knows when it should be decoding or presenting a particular picture or audio sequence. By comparing it with its internal clock (kept updated by SCR and PCR), the decoder 'knows' when to output a given entertainment 'chunk'. The decoder either skips or repeats picture displays to ensure that the PTS is within one picture's worth of 90 kHz clock 'ticks' of the SCR when a picture is displayed. If the PTS is earlier (has a smaller value) than the current SCR, the decoder discards the picture. If the PTS is later (has a larger value) than the current SCR, the decoder repeats the display of the picture.

Splicing bitstreams

As we already saw in Chapter 8, an MPEG encoded picture can only be 'cut' at an I-frame boundary or, in the language of digital television, at a *random-access* point. Clearly the transport stream can only be cut if provision is made in the PES for a convenient cut-point to be found. A random entry point in the transport stream is indicated by a flag in the adaptation header of the packet which contains a random-access point for the elementary bitstream. In addition, the splice_countdown field may be included as part of the adaption header. This indicates the number of packets with the same PID as the current packet that remain in the bitstream until a splicing point packet. It thus pre-signals the point at which a switch from one program segment to another may occur. At the transport stream level, PES packets are arranged into the packetized transport stream by means of the variable length adaption header, so that the random access points in the PES come at the start of the transport level packets. This greatly simplifies the switching of DTV bitstreams and results in the efficient (i.e. quick) re-establishment of picture after the switch point.

Conditional access table

Along with the PAT and the PMT there exists one other extremely important table on the multiplex. This is the conditional access table (CAT), which is transmitted at PID = 1 when one programme or more on the multiplex is scrambled. Importantly, the CAT does not carry entitlement information but information about the management of entitlement information. Conditional access is covered later in the chapter.

DVB service information

To the three tables already described, which are a part of the MPEG standard and are termed PSI or programme-specific information, the European DVB project has added its own tables to the MPEG-II multiplex. These are largely aimed at making the choice of television programmes on a particular multiplex – or number of multiplexes – more user friendly and interesting. It is the DVB service information (along with the PSI) that is used to build the electronic programme guide (EPG); the real-time *Radio Times* of digital television!

DVB-SI tables include:

- the network information table (NIT), which specifies other RF channels that contain associated, but different multiplexes; for instance, other multiplexes operated by the same television company
- the service description table (SDT), which lists parameters associated with each service on the multiplex

- the event information table (EIT), which contains information on programme timings
- the time and date table (TDT), which is used to update the clock and calendar in the set-top box should this wander from the real time.

DVB have also suggested optional tables specifying groups of services ('bouquets'), which present viewing options to the viewer in a more amenable (tempting!) way.

Conditional access

Digital television has arrived because large organizations believe they can make a great deal of money. Central to their plans is the idea that television will become more and more 'targeted', with programming to suit interest groups who will be prepared to pay to watch the programmes they like. This requires a system of scrambling and of conditional access so programmes are not receivable unless the viewer has paid for the privilege. Note that these two components, the scrambling and the conditional access, are separate things and, although the DVB system in Europe specifies the scrambling and descrambling approach to be used (common scrambling algorithm), it does not specify the conditional access system or the descrambling keys and how these are obtained at the decoder. This 'private' nature of conditional access ensures that one network operator cannot hack into the user database of another, and results in the rather complicated situation that a particular programme on a multiplex may have several different conditional access systems. For instance, a sci-fi channel may be obtainable on the same multiplex from several different operators, each with their own conditional access system. This is the reason for the CAT (conditional access table), which specifies the PIDs for the packets containing the entitlement management messages (EMMs) for each conditional access system. Information in the link header of a transport packet indicates whether the payload in the packet is scrambled and, if so, flags the key to be used for descrambling. The header information in a packet is always transmitted 'in the clear', i.e. unscrambled. Scrambling may be performed at the PES or the transport level. Clearly, information on either scrambling systems or conditional access mechanisms is highly commercially sensitive. For this reason only the most sketchy of information is available, except to legitimate operators under non-disclosure agreements – so-called 'custodians'.

SimulCrypt and MultiCrypt

In order that the digital receiver can descramble programmes which have been scrambled by different conditional access systems, a common

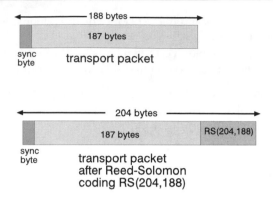

Figure 9.2   *MPEG data packet structure*

interface (CI) has been defined and can be built into the digital receiver or set-top box (as we shall see in Chapter 11). Based on an array of computer-style PCMCIA modules, different CA systems can be addressed sequentially by the integrated receiver decoder (IRD), each module decrypting one or more programmes on the multiplex. The term Multi-Crypt is used to describe the simultaneous operation of several CA systems. The alternative is SimulCrypt. In this case, commercial negotiations between different programme providers have led to a contract which enables viewers to use the one specific CA system built into their IRD (or on one PCMCIA CI module) to watch all the programmes that were scrambled under the aegis of several CA systems. SimulCrypt relies on various operators working together under a code of conduct, the reality of which is yet to be proved!

Channel coding

The upper part of Figure 9.2 illustrates the MPEG system packet of 188 bytes. To achieve the appropriate level of error protection required for cable transmission of digital data over long distances, several coding techniques are employed one after the other. These consist of randomization for spectrum shaping (or scrambling), Reed-Solomon encoding and convolutional interleaving. These techniques extend the packet length to 204 bytes overall. It's important to bear in mind that both forms, the un-coded 188-byte packet and the encoded 204-byte packet, are found in practical implementations of the MPEG interface.

Randomization (scrambling)

Although not strictly a coding technique, in randomization the system input stream is first randomized. This has two effects; it reduces DC

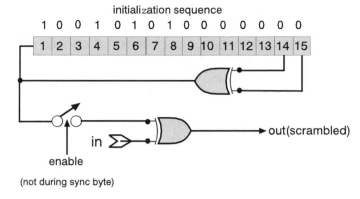

Figure 9.3 *Scrambler or randomizer*

content, which means the signal can be reliably AC coupled (by a transformer for example), and it ensures there are adequate binary transitions for reliable clock recovery in the decoder. The polynomial for the pseudo random binary sequence (PRBS) generator is:

$$1 + X^{14} + X^{15}$$

and this is implemented in hardware as illustrated in Figure 9.3. The PRBS registers must be loaded with the initialization sequence 100101010000000 at the start of every eighth transport packet, and the receiver must be able to determine this sequence. To this end, the action of the 'scrambler' is keyed so that sync bytes are not scrambled. Furthermore, the value of every eighth sync_byte is bitwise inverted from 47 h to B8 h.

Reed-Solomon encoding

Following the scrambling process described above, a systematic Reed-Solomon encoding operation is performed on each randomized MPEG-II transport packet. Reed-Solomon encoding and decoding is based on a specialist area of mathematics known as Galois fields or finite fields. These arithmetic operations require special hardware or software functions to implement, but the Reed-Solomon code is a very powerful coding technique and is widely used in disk and tape data recording as well as in MODEM and broadcast applications. This is due to the fact that Reed-Solomon codes are particularly well suited to correcting burst errors, where a series of bits in the code word are received by mistake – such as may be experienced in disk/tape dropout or in radio reception.

A Reed-Solomon code is specified as RS(n, k) with s bit symbols. This means that the encoder takes k data symbols of s bits each and adds parity symbols to make an n symbol codeword. There are therefore $n - k$ parity symbols of s bits each. A Reed-Solomon decoder can correct up to t symbols that contain errors in a code word, where $2t = n - k$. The amount of computation required to encode and decode Reed-Solomon codes is related to the number of parity symbols per code word. A large value of t means that a large number of errors can be corrected, but this requires more computational power than a small value of t. Reed-Solomon algebraic decoding procedures can correct errors and erasures. An erasure occurs when the position of a corrupted symbol is known. A decoder can correct up to t errors or up to $2t$ erasures. Erasure information can often be supplied by the demodulator in a digital communication system, because the demodulator may be able to 'flag' received symbols that contain errors.

In this process, an MPEG transport stream of 188 bytes has 16 parity bytes added to it to make a total of 204 byte code word or RS(204, 188), $t = 8$, as illustrated in the lower part of Figure 9.2. (Note that RS coding leaves the original 188-byte data as it was; it simply adds the parity information afterwards. In this way the packet sync_byte values are preserved.)

Convolutional interleaving

Reed-Solomon encoding is very powerful, but the error correction capacity is limited to the packet timeframe. Moreover, in the case of a bad burst error, when a great number of bytes are damaged, it's quite possible that too much disruption will have occurred within one single frame to recover the original data. If, however, the data is interleaved, a bad burst error of contiguous interleaved data will be less harmfully distributed among several frames when a reverse de-interleaving process is performed in decoding.

The interleaving approach used in DTV is termed a Forney interleave. The interleaver is based in 12 branches, as illustrated in Figure 9.4. Each branch has a different delay, and the switches feeding the signal in and out

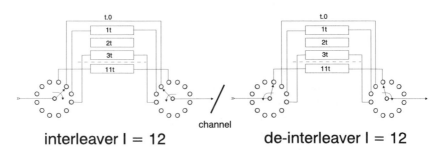

Figure 9.4   *Forney interleaver*

of the delay branches advance at 1-byte intervals. This process therefore 'spreads' the input data in time, but in a predictable way. The receiver has to perform the inverse process, and this too is illustrated in Figure 9.4.

Standard electrical interfaces for the MPEG-II transport stream

As we have seen, a transport stream is made up of packetized elements known as transport packets. The packets are either 188 bytes long or 204 bytes long if they are Reed-Solomon encoded. The electrical interfaces do not specify if the data has (or has not) to have Reed-Soloman encoding; this is optional. Each MPEG-II transport packet looks like one or the other of the two packets illustrated in Figure 9.2, depending whether Reed-Soloman coding has been used or not. There are three different interfaces that are specified to carry MPEG-II data over various distances; the synchronous parallel interface (SPI), the synchronous serial interface (SSI) and the asynchronous serial interface (ASI).

Synchronous parallel interface

The synchronous parallel interface (SPI) is implemented as shown in Figure 9.5. It is designed to cover short to medium distances, rather like the old 601 parallel interface. The data and clock signals are obvious enough, but this interface has two further signals, as shown. PSYNC is used to identify the beginning of each data packet, and DVALID is used to flag empty (and therefore non-valid) bytes that occasionally appear in a

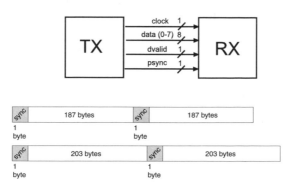

Figure 9.5 *Synchronous parallel interface*

Table 9.1   *Electrical characteristics of the SPI*

Source characteristics

Output Z	100 ohms
Common mode volts	1.125 – 1.375 V
Signal amplitude	247 – 454 mV

Destination characteristics

Input Z	90 – 132 ohms
Max input signal	2.0 V pk–pk
Minimum input signal	100 mV pk–pk

Table 9.2   *Pinout specification for the SPI*

Pin	Signal	Pin	Signal
1	Clock A	14	Clock B
2	Gnd	15	Gnd
3	Bit 7A	16	Bit 7B
4	Bit 6A	17	Bit 6B
5	Bit 5A	18	Bit 5B
6	Bit 4A	19	Bit 4B
7	Bit 3A	20	Bit 3B
8	Bit 2A	21	Bit 2B
9	Bit 1A	22	Bit 1B
10	Bit 0A	23	Bit 0B
11	DVALID A	24	DVALID B
12	PSYNC A	25	PSYNC B
13	Cable shield		

non-Reed-Solomon encoded 204-byte version of the bitstream. In this implementation the last 16 bytes of the 204-byte packet are not parity bytes but just 'dummy' bytes, and therefore have to be signalled as such. The clock signal corresponds to the useful bit-rate, and never exceeds 13.5 MHz.

The electrical characteristics of the interface are defined in Table 9.1. Mechanically, the interface uses a D-25 plug and socket arrangement with the pinout specified in Table 9.2.

Synchronous serial interface

The synchronous serial interface (SSI) can be seen as an extension of the parallel interface. This interface has no fixed rate; instead, the data is directly converted to serial from the parallel interface and is bi-phase encoded before being buffered for physical interfacing. There is an alternative that specifies a fibre-optic interface. Despite the standard, SSI is not widely implemented, and the interested reader is referred to the standard document (EN 50083-9: 1997) for further information.

The asynchronous serial interface

The asynchronous serial interface (ASI) is the most widely implemented electrical interface for MPEG-II coded signals. Unlike the SSI, the asynchronous interface guarantees a fixed data rate irrespective of the data-rate of the transport packets comprising the transport stream. This is particularly useful from a practical point of view because phase-locked loop receiver technology can be used to extract clock information, enabling the interface to be used over longer distances and in more noisy environments. ASI is derived from the basic electrical (and optical) and coding format of the computer industry Fibre Channel interface. In spite of its name, this interface exists on fibre and on cable, as does the ASI.

Like the other two interfaces, ASI is a point-to-point interface, not a network. Figure 9.6 illustrates implementation of the ASI over 'copper' (i.e. coaxial cable) and over fibre. The 8-bit data is first turned into 10-bit data in a process called 8B/10B coding; in other words, each 8 bit word is used to look up a 10 bit result in a look-up table. These 10-bit codes are used to ensure a lack of DC content and also to enable the use of specific codes for sync words. These 10-bit words are then passed through a parallel-to-serial converter, which operates at a fixed output rate of 270 Mbits/s. If there is not sufficient data at the input of the parallel-to-serial conversion stage, the converter substitutes synchronization words in place of data.

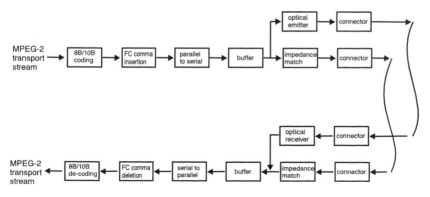

Figure 9.6 *Asynchronous serial interface*

Table 9.3   *Electrical characteristics of the ASI*

Source characteristics	
Output voltage	800 mV
Max. rise/fall time	1.2 nS
Receiver characteristics	
Min. input voltage	200 mV
Max. input voltage	880 mV
Return loss	15 dB (0.3 MHz–1 GHz)

The interface is thereby 'stuffed' to ensure a continuous data rate. The data then passes over a fibre or coaxial link to the receiver.

In the receiver, the serial data bits are recovered from a 'flywheeled' PLL circuit and passed to the decoder, which functions just like an encoder in reverse. In order to recover byte alignment, a special sync byte (Fibre Channel comma or FC comma) is inserted by the coder and recovered by the receiver in order to indicate the boundary of bytes within the interface.

Electrical medium characteristics are designed to be similar to the serial digital video interface (SDV) in that the signal is designed to be carried on coaxial cable terminated with BNC connectors. The electrical characteristics of the signal are given in Table 9.3. Importantly (and here the interface does not match its video cousin), the standard specifies the use of a transformer for coupling the signal to the cable.

Fibre optic implementations are designed to be a 62.5-μm core multimode fibre interfaced on SC (IEC 874-14) connectors using LED transmitter in the 1280-nm wavelength region.

10
Broadcasting digital video

Digital modulation

Information theory tells us that the capacity of a channel is its bandwidth multiplied by its dynamic range. Clearly the simplest method of transmitting digital binary data is by a simple amplitude modulation scheme, like that shown in Figure 10.1a. This technique is known as amplitude shift keying (ASK). This isn't very efficient in bandwidth terms because, if we take the available channel bandwidth to be F, it's only possible to achieve in the region of $2F$ bits/s as a maximum data rate. (Note that it's important to 'shape' the digital pulses in order to limit the sideband energy, which would be very great with fast-edged pulses.) Furthermore, it isn't a very efficient usage of the analogue channel, which may have a signal-to-noise ratio much greater than is strictly necessary for binary transmission. Channel capacity can therefore be improved by the use of a multiple level digital system, like that shown in Figure 10.1b. In this example, four amplitude states are used to signal; 00, 01, 10, 11. This increases the channel capacity by twice; say $4F$ bits/s. If we use eight levels, the channel capacity improves to $8F$ bits/s. This latter scheme (with eight levels) is essentially the system adopted by the ATSC for the American led DTV development. This is how they manage to achieve a 19.28 Mbits/s payload in a 6 MHz channel. The ATSC standard also provides for a 16-level system for use in good reception conditions (signal to noise ratio better than 28 dB), which increases payload to 38.57 Mbits/s.

Quadrature amplitude modulation

Instead of the simple diode demodulator shown in Figure 10.1 to demodulate the amplitude modulated carrier, synchronous techniques are almost always employed. This leads to the possibility of using a technique similar to the one we saw in relation to the modulation of

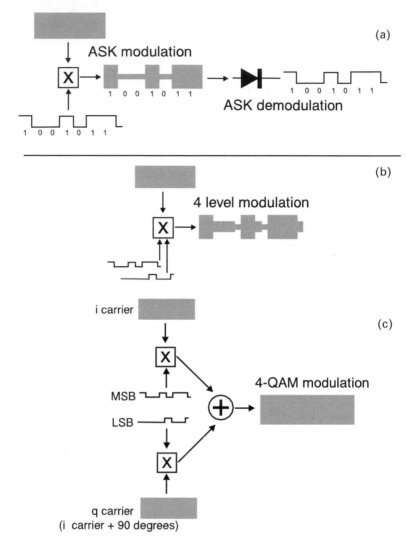

Figure 10.1 (a and b) *Amplitude-shift keying (ASK), multilevel version at (b); (c) Quadrature amplitude modulation (QAM) process*

colour information in NTSC colour system. In this case, two carriers (i and q) are used, where q (quadrature) is 90 degrees phase shifted in relation to i (in-phase) (see Figure 10.1c; compare with Figures 2.13 and 2.14 in Chapter 2).

An NTSC signal is modulated by a continuous analogue colour signal. In a digital system, two quadrature carriers are modulated by a digital signal of a limited number of pre-defined levels. This leads to a limited

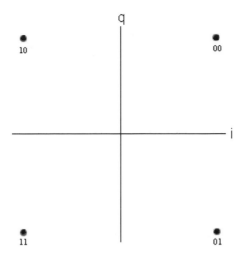

Figure 10.2   *4-state QAM or quadrature phase-shift keying (QPSK)*

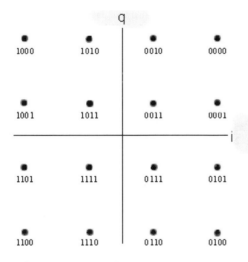

Figure 10.3   *QAM-16 constellation*

number of possible carrier states, which may be illustrated as in Figures 10.2 to 10.4. The gamut of possible modulation states is termed a 'constellation', the number of possible constellation states being appended to the term QAM (quadrature amplitude modulation) to define the system in use – for instance, QAM-4 (also known as quadrature phase shift keying, or QPSK), QAM-16 and QAM-64, all of which are illustrated. Also look at Figure 10.5, which illustrates the results of a real

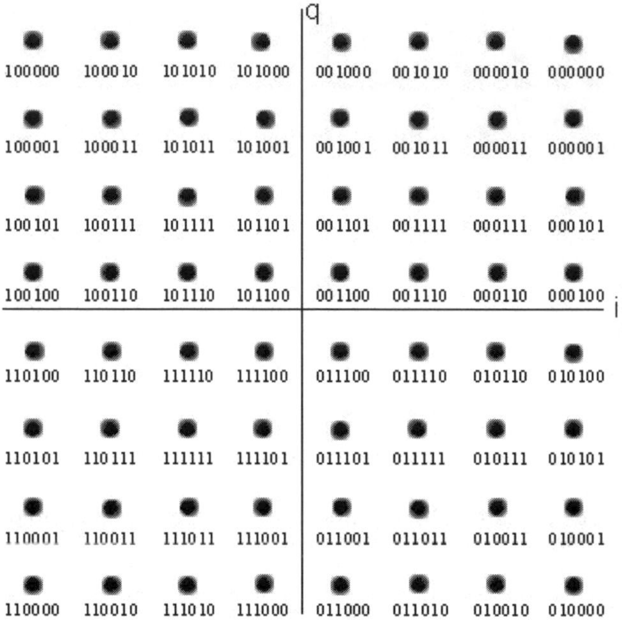
q
i
100000 100010 101010 101000 001000 001010 000010 000000
100001 100011 101011 101001 001001 001011 000011 000001
100101 100111 101111 101101 001101 001111 000111 000101
100100 100110 101110 101100 001100 001110 000110 000100
110100 110110 111110 111100 011100 011110 010110 010100
110101 110111 111111 111101 011101 011111 010111 010101
110001 110011 111011 111001 011001 011011 010011 010001
110000 110010 111010 111000 011000 011010 010010 010000

Figure 10.4 *QAM-64 constellation*

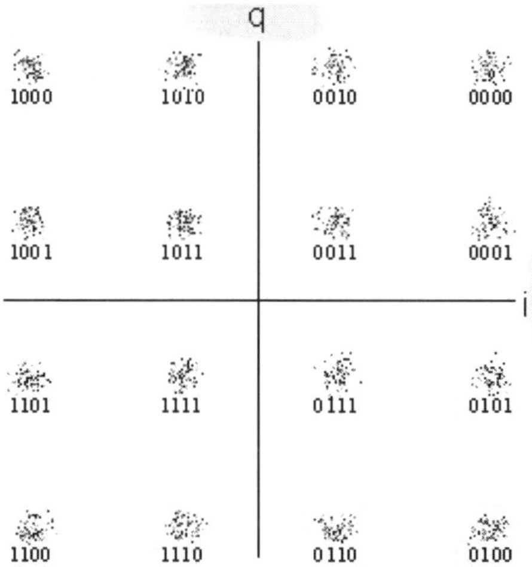
q
i
1000 1010 0010 0000
1001 1011 0011 0001
1101 1111 0111 0101
1100 1110 0110 0100

Figure 10.5 *Result of practical QAM-16 code and decode process with noise* demodulation process in which the individual points within the constellation are subject to the effects of noise. Clearly there lies a practical noise limit where the 'spread' of the individual carrier states is sufficiently great that it is no longer possible to determine the coded value with certainty. This noise limit will obviously be greater the fewer the number of possible carrier states and their consequently greater separation.

This explains why there are a multiplicity of different DTV modulation techniques. If you think of channel capacity as a physical cross-section of a rectangular data 'pipe', you can imagine bandwidth as width and dynamic range as depth (i.e. the vertical dimension). A given television multiplex signal will require, one way or another, the same cross-section of data pipe but – given the prevailing circumstances – the data pipe may be of different shapes. Essentially, each modulation technique is an attempt to balance the requirements of bandwidth and dynamic range in order to achieve the best channel capacity for a given channel. Digital television by satellite, by cable and by terrestrial transmission systems have all struck different balances.

Modulation for satellite and cable systems

Current digital satellite television standards are designed for use with transponder bandwidths of 26–72 MHz, which encompass the vast majority of current Ku-Band TV satellites, such as the Astra and Hotbird clusters. Digital television transmission reflects many of the same choices made in analogue television modulation systems. For instance, satellite television broadcasting has always used a much wider channel than that used for terrestrial or cable broadcasts. Satellite broadcast reception is characterized by a very poor signal-to-noise ratio, perhaps as low as 10 dB! For analogue systems, FM modulation is necessary because it is robust to amplitude distortion and noise due to the limiting effect in the IF stage prior to demodulation. It is for this reason that a wider bandwidth is required to cope with the much more complicated sideband structure of an FM signal. It is therefore no surprise to find that digital satellite transmissions use QPSK, i.e. the use of high-rate symbols of relatively few levels. Using our data pipe analogy, we can see that a satellite multiplex channel is very wide but not very deep.

Cable provides a relatively benign transmission environment, with low noise and relatively low reflections. This implies that the channel bandwidth can be commensurately reduced and more signal amplitude levels used, and indeed that is the case. So, for digital transmission by cable, a 16-QAM or even 64-QAM system may be used. Table 10.1 tabulates the main characteristics of cable and satellite transmission techniques.

Table 10.1   *The main characteristics of cable and satellite transmission techniques*

Parameter	Satellite	Cable
Channel width	26–54 MHz	8 MHz
Modulation type	QPSK	64-, 32- or 16-QAM

Establishing reference phase

QAM (like NTSC and PAL) could provide a period of unmodulated carrier in order to establish a reference phase. Alternatively a differential technique may be employed, as in the modulation of the NICAM carrier for stereo digital sound in analogue television, discussed in Chapter 2. The technique involves coding a change in phase state to represent a value; for instance 90° change = 00, −90° = 01, and so on. In this way, the need for an absolute reference phase is avoided. This modulation technique is known as differential quadrature phase shift keying (DQPSK). QPSK systems for digital television by satellite could use a DQPSK approach, but instead an absolute phase modulation system is employed. However, when demodulated, only one of the four possible phases produces valid framing data. The receiver thereby 'seeks out' this valid phase in the process of capturing a new digital signal by trying each of the four possible states in turn until valid data is extracted.

Modulation schemes for digital television by cable use differential coding of the MSBs in the QAM-16 (or QAM-64) system in order to avoid the requirement for a reference phase period. Note that only the MSBs need to be differentially coded in order to establish the quadrant.

Convolutional or Viterbi coding

We have already discussed error protection coding in the last chapter; this included both Reed-Solomon encoding and convolutional interleaving. In digital television, these processes are termed outer coding. There is, however, yet another form of error protecting coding, which is performed prior to modulation for transmission by satellite, cable or terrestrial channels (although the details differ slightly). This third regime, termed inner coding, is known as convolutional or Viterbi coding. The process of Viterbi coding is illustrated in Figure 10.6. The input data stream passes down a chain of shift-registers, the outputs of which are modulo summed (or, in some cases, not summed) with the incoming serial data. This process produces, in its purest state, two bitstreams in place of one, and at the same rate as the original input data-rate. Not surprisingly, with a

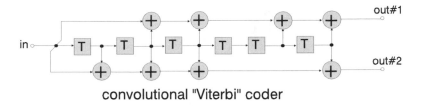

convolutional "Viterbi" coder

Figure 10.6 *Viterbi coding*

redundancy of 100 per cent, this provides a powerful correction capacity. The corresponding decoding algorithm (due to the eponymous Viterbi) involves using a probability approach to derive the likely input data even in the presence of very considerable numbers of errors. In the case of satellite modulation, the two outputs of the Viterbi coder can be used to derive the I and Q inputs to the QPSK modulator directly. For other applications (cable and terrestrial) the outputs can be serialized, some-times at a rate lower than twice the input rate, by deliberately not taking every bit. In this case the coding is known as 'punctured' Viterbi coding.

Terrestrial transmission – DVB-T (COFDM) and US ATSC (8-VSB) systems

Terrestrial digital television broadcasting enjoys a much greater signal-to-noise ratio than does transmission by satellite; a fortunate circumstance, because each channel multiplex has only been allocated the much lower 6–8 MHz bandwidth of an existing analogue channel. To use our analogy of data pipe cross-section again, the data pipe for terrestrial broadcasting is much less wide than the satellite pipe, but much deeper. This implies the inevitability of a multi-state modulation scheme like 64-QAM, or a multi-state amplitude modulation system. This latter approach is, in fact, the one taken by the engineering consortium which developed the modulation scheme for terrestrial transmission in the United States (the ATSC). However, terrestrial television is beset by multi-path distortion effects, as illustrated in Figure 10.7. Multi-path is an irritating feature of analogue systems, where it produces 'ghosting' (as illustrated in Figure 10.7), but in a digital system the echoes can blur individual digital symbols (1s and 0s) together, making it impossible to decode. These multi-path effects have presented a virtually unprecedented challenge to the designers of digital TV terrestrial transmission systems, ruling out, as they have done, the use of the straightforward modulation systems described above. The two engineering approaches taken to achieve a practical realization of digital terrestrial TV transmission are interesting. In America the approach has been evolutionary, empirical and has resulted

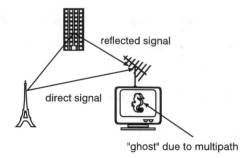

Figure 10.7   *The effect of multipath on an analogue television signal*

in a simple modulation scheme being appended with considerable complexity. In Europe the approach was revolutionary, theoretical and has resulted in a complex system of extreme elegance. This modulation system is known as coded orthogonal frequency division multiplex (COFDM).

Coded orthogonal frequency division multiplexing (COFDM)

Orthogonal frequency division multiplexing (OFDM) is a multi-carrier transmission technique that divides the available spectrum into many carriers (over 2000 in the UK), each one being modulated by a low-rate data stream. In OFDM, the spectrum is used very efficiently by spacing the channels very close together and preventing interference between the closely spaced carriers, each of which is *orthogonal* to the others. The orthogonality of the carriers means that the spectrum of each modulated carrier is arranged so that it has a null at the centre frequency of each of the other carriers in the system (Figure 10.8). This results in no interference between the carriers, allowing then to be spaced as close as theoretically possible. Each carrier in an OFDM signal has a very narrow

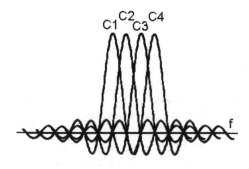

Figure 10.8   *Adjacent OFDM carriers*

bandwidth (1 kHz); thus the resulting symbol rate – per carrier – is very low. This results in the signal having a high tolerance to multi-path delay, as the delay spread must be very long to cause significant inter-symbol interference (typically > 500 μs). Nevertheless, OFDM can be prone to errors due to frequency selective fading, channel noise and other propagation effects, which is why COFDM is often employed. COFDM is the same as OFDM, except that forward error correction is applied to the signal before transmission. In digital terrestrial television, the forward coding is the Viterbi type coding discussed above.

Practical COFDM

It will not surprise you to know that COFDM is not achieved practically using thousands of oscillators and modulators! In fact, the signal is generated in the frequency domain and transformed back into the time domain for transmission. At the receiver, a Fourier transform for transmission (FFT) is performed in order to extract each carrier: Each carrier is assigned some data to transmit. The required amplitude and phase of the carrier is then calculated based on the modulation scheme (typically differential QPSK or QAM), and the required spectrum is then converted back to its time domain signal using an inverse Fourier transform for transmission. Figure 10.9 illustrates a basic OFDM transmitter and receiver.

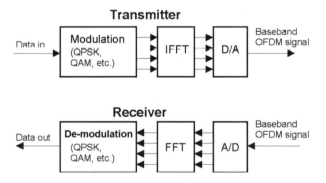

Figure 10.9 *COFDM code-decode process*

Adding a guard period to OFDM modulation

One of the most important properties of OFDM transmissions is its resistance to multi-path distortion effects; this is achieved by having a long symbol period, which minimizes the inter-symbol interference. The level of robustness can be increased even more by the addition of a guard period between transmitted symbols. The guard period allows time for multi-path signals from the previous symbol to die away before the information from the current symbol is recovered. As long as the multi-path delay echoes stay within the guard period duration, there is strictly no limitation regarding the signal level of the echoes; they can even exceed the signal level of the direct path! Multi-path echoes delayed longer than the guard period will have been reflected off very distant objects, and so will not cause inter-symbol interference.

The advantages of COFDM

Although COFDM modulation is *not* achieved with 2000 oscillators and modulators, its remarkable properties can be understood by suspending disbelief for a moment and imagining that the system does indeed work in this fashion. Imagine 2000 oscillator circuits each being ASK modulated by a low-rate signal of a few kbits per second. Two thousand carriers in an 8-MHz channel suggests a spacing of 4 kHz per carrier because

$$8\,000\,000/2000 = 4000$$

Provided the modulation frequencies are low enough (say 1 kHz for 2 kbits/s), spacing like this should be sufficient to prevent the sidebands from one carrier interfering with the sidebands of the next. (It is the precise calculation of spacing and keying frequency that accomplishes the 'or-thogonality' alluded to earlier.) Thus we can think of a COFDM television signal as one TV signal multiplex broken into 2000 low bit-rate (low bandwidth) radio stations, each spaced at 4 kHz. Multi-path distortion disrupting a few carriers will thereby only destroy a small part of the overall data transmission bandwidth, which can be recovered with the use of appropriate error correcting coding. Further immunity from multi-path can be won by keying the carrier for only a fraction of the given symbol period (which we saw is termed as adding a guard-band). In this way, multi-path 'echo' can be 'windowed out' in the reception process. COFDM even provides the possibility for mobile television reception because, even if multi-path conditions continually change, provided enough data is received from the undamaged carriers correct reception will result. CODFM's further advantage is that, because interference from another transmitter is really only a severe form of multi-path, its inherent resistance to this type of distortion means that transmitter networks can operate at the same frequency for a given single multiplex. Thus, single-frequency network digital television may remove, at a stroke, the huge frequency spectrum-planning task that analogue television required where repeaters and transmitters had to use a band of frequencies to avoid co-channel interference for viewers in fringe areas. (Actually, the reality of single-frequency networks increases with the number of OFDM carriers. The DVB Committee originally defined an OFDM system with 8000 carriers, but this was reduced to 2000 for UK because of doubts as to whether silicon vendors could produce chips to perform an 8000 FFT at a reasonable cost for a set-top box. In fact they could, and the 2000 carrier system in the UK does pose some practical limits to single frequency networks.)

Many explanations of COFDM are very mathematical. The advantage of the '2000 radio stations' explanation is that it is simple but retains the worthy impression of the ingenuity of this modulation process. COFDM's only disadvantage is integrated-circuit complexity. This consideration alone led engineers in the USA to develop the 8-VSB system which, as we have already seen, is a multi-level coded amplitude modulation scheme. However, in order to make this scheme practical, the engineers have been forced to append considerable complexity.

8-VSB modulation

The 8-VSB single-carrier modulation system adopted in America delivers 19.29 Mbits/s in a 6 MHz terrestrial channel. The serial data stream comprises the now familiar 188-byte MPEG-compatible data packets. Following randomization and forward error correction processing, the data packets are formatted into 'data frames' for transmission and a data 'segment sync' and a 'data field sync' are added. Rather like analogue TV, each data frame consists of two data fields, each containing 313 data segments. Each data segment is about 77.7 µS – not far from analogue TV line frequency! And each has a variable, signal portion and a sync portion as illustrated in Figure 10.10. This is easier to see when the segment is

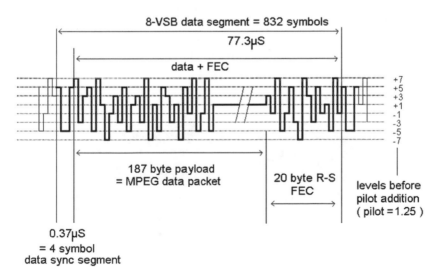

Figure 10.10 *8-VSB data segment*

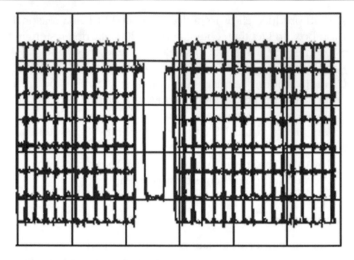

Figure 10.11   *Modulated 8-VSB signal*

viewed as on an oscilloscope (Figure 10.11). With the static sync part and the modulated data part, at approximately the same line-frequency as analogue TV, the evolutionary nature (rather then the revolutionary nature) of 8-VSB may be seen. When viewed as data frames and fields (Figure 10.12), the similarity is even more evident!

The first data segment of each data field is a synchronizing signal, which includes the training sequence used by the equalizer in the receiver. The remaining 312 data segments each carry the equivalent of the data from one 188-byte transport packet plus the forward error correction overhead. Each data segment consists of 832 symbols. The first four symbols are transmitted in binary form and provide segment synchronization. This data segment sync signal also represents the sync byte of the 188-byte MPEG-II

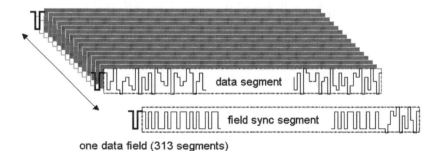

one data field (313 segments)

Figure 10.12   *The 8-VSB data field*

compatible transport packet. The remaining 828 symbols of each data segment carry data equivalent to the remaining 187 bytes of a transport packet and its associated error correction overhead. These 828 symbols are transmitted as 8-level signals and therefore carry three bits per symbol. The symbol rate is 10.76 Msymbols/s and the data frame rate is 20.66 frames/s (once again, very much like analogue TV in nature). To assist receiver operation a pilot-carrier is included at approximately 310 kHz from the lower band edge.

Like COFDM, one of the requirements of the 8-VSB system is that it must work in the presence of multi-path interference. Since 8-VSB is an entirely time-domain coding scheme, the approach used is an adaptive 'echo-cancelling' equalizer. As mentioned earlier, each field data segement contains a robust two-level signal which is used for equalizer 'training'. That's to say that the relections from this training-sequence are used, within a feedback loop, to adjust the phase-correcting equalizer and remove multi-path echos. It is in this area that the 8-VSB system has proved to be rather poor and has given engineers some of the most difficult challenges.

16-VSB

The field data segments also contain the mode signalling to indicate whether the eight level (8-VSB) system is being used, or the higher data-rate, 16 level (16-VSB) system. This alternative modulation scheme is intended for less stringent environments (with relatively little or no multipath) and carries twice the data-rate of 8-VSB. The data segment sync and frame sync are essentially the same as for 8-VSB.

Hierarchical modulation

The following paragraph illustrates the principle of hierarchical modulation, in that some symbols are transmitted with a higher priority than others and yet are used to convey the basic information, albeit in a simplified form.

TODAY there are very STRONG, icy, northerly WINDS forecast, starting from about midday. These may contain GUSTS up TO about 100 KM per HOUR. These will be accompanied by driving rain which will result in POOR VISIBILITY.

We can imagine that the capitalized symbols could be sent more frequently or using a more complex data-protection and correction scheme. In this way, in conditions of poor reception, even if the majority of the message were lost, the salient points would still be received and understood.

In the context of television, hierarchical modulation can be used to send a high-definition picture, and a lower standard-definition picture simultaneously, so that – in conditions of poor reception – the receiver could revert to a standard definition picture rather than simply fail to produce an HD image. In the context of, for example 8-VSB modulation, the symbols for the lower definition picture (our capitalized symbols above), would be restricted to perhaps four, or even two levels, the highest and the lowest for example. Thereby, even in extremely noisy conditions, the symbols could be received and decoded, even when the more complex modulation scheme was completely defeated by noise.

QAM modulation schemes too can operate effective hierarchical modulation schemes by restricting the priority symbols to a limited gamut of the possible constellation of modulation states. It's quite straightforward to imagine, for example, a QAM-4 system, 'riding on' a QAM-16 signal, with the priority signals occupying generalized positions in each of the four modulation quadrants.

Interoperability

Interoperability refers to the ability of the MPEG-II digital television multiplex to be 're-packed' into other transport systems; notably the national and international telecommunication networks. As we have already seen, packets can be broken down into smaller packets or 'stuffed' together to form bigger packets, provided enough associated information is contained in pre-defined places for the packets to be 'unpacked' at the end of their journey. On the face of it, it may seem as if there is nothing that prevents the transmission of a bitstream as the payload of a different transmission system. It may be complicated and fiddly, but it should always be possible. However, interoperability has two aspects; the first is simply the mapping of the digital television information into another data structure as stated above, and the second relates to the delivery of the bitstream in real time. That is to say, the output bitstream of the alternative transport system must have the proper real-time characteristics; the data mustn't 'choke' somewhere in the system, causing the picture to freeze!

Interoperability with ATM

Because ATM is expected to form the basis of future broadband communications networks, the issue of bitstream interoperability with ATM networks is especially important. Happily, the MPEG-II transport packet size is such that it can be easily partitioned for transfer in a link layer that supports asynchronous transfer mode (ATM) transmission.

ATM cell and transport packet structures

The ATM cell consists of two parts; a 5-byte header and a 48-byte information field. The header, primarily significant for networking purposes, consists of the fields shown in Table 10.2.

The ATM user data field consists of 48 bytes, where up to 4 bytes can be allocated to an adaptation layer.

The MPEG-II transport layer and the ATM layer serve different functions in a video delivery application. The MPEG-II transport layer solves MPEG-II presentation problems, and performs the multimedia multiplexing function. The ATM layer solves switching and network adaptation problems.

Figure 10.13 illustrates one of several possible methods for mapping the MPEG-II transport packet into the ATM format.

Table 10.2 *ATM cell header fields*

GFC	A 4-bit generic flow control field used to control the flow of traffic across the user network interface (UNI); exact mechanisms for flow control are under investigation
VPI	An 8-bit network virtual path identifier
VCI	A 16-bit network virtual circuit identifier
PT	A 3-bit payload type (i.e. user information type ID)
CLP	A 1-bit cell loss priority flag (eligibility of the cell for discard by the network under congested conditions)
HEC	An 8-bit header error control field for ATM header error correction
AAL	ATM adaptation layer bytes (user specific header)

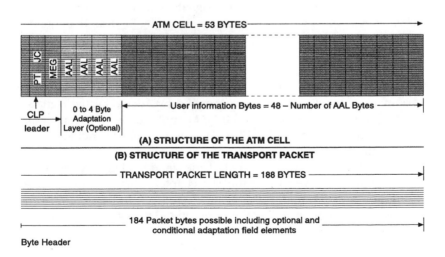

Figure 10.13 *Mapping video into ATM4*

11
Consumer digital technology

Receiver technology

Irrespective of whether a DTV signal issues from a terrestrial transmitter or a satellite, or arrives by cable, many of the functions of a digital receiver are common. As we have seen, the source coding used is always MPEG-II compression in Europe and MPEG-II video with AC-3 audio compression in the US. The differences in the decoders, in equipment designed for the various delivery formats, is the corollary of the different modulation technologies we studied in the previous chapter. Notably, whereas satellite modulation is QPSK, cable is n-QAM, and terrestrial is COFDM or 8-VSB.

Looking at Figure 11.1, let's consider the jobs that a digital receiver must perform in order to display a digital television programme:

1 It must amplify the incoming RF signal; tune to the correct digital channel and thereby down-convert the incoming RF to an IF frequency. In the case of satellite, the signal is already down converted once inside to low-noise converter (LNC) or low-noise block (LNB) at the antenna head.

2 It must demodulate the IF into a base-band bitstream. This process may be analogue or digital, and produces the two demodulated signals (I and Q) in the case of QPSK and n-QAM applications, or a single multi-level output in the case of the 8-VSM system. Following demodulation, the ensuing processes must now occur:

- Viterbi decoding – the reverse of the 'inner-coding' process of the procedure that we saw in the last chapter. This produces a single bitstream, which must be,
- Forney convolutional de-interleaved, in order to get all the bits back into the right order again (see Chapter 9), and
- Reed-Solomon decoded – to get back to 188-byte MPEG-II transport packets. It is at this point that one or more PCMCIA modules may

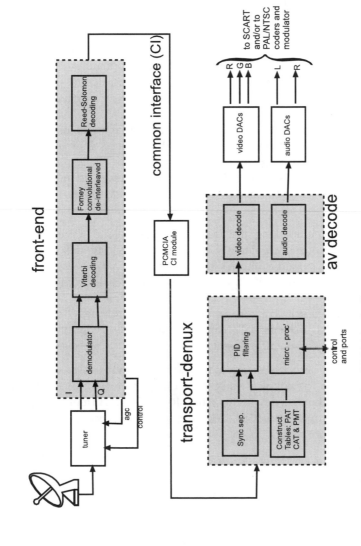

Figure 11.1 Integrated receiver decoder (IRD) block-diagram

intercept the raw parallel-transport stream via the standardized interface known as the common interface (CI).

3 The signal now has to be de-multiplexed, and this involves the process termed PID filtering. This embodies:

- 'Sync-ing' the decoder by looking for the first byte in the transport-packet header,
- Finding the PID = 0 (see Chapter 9) and using this to construct the program allocation table (PAT), and finding PID = 1 to construct the conditional access table (CAT), if appropriate.
- Using the information in the PAT to construct the program map table (PMT) and the CAT and provide the viewer with a choice of available programmes on the multiplex.

4 The viewer will now have to provide some input (usually via a remote control module that converses with the receiver's microprocessor), and will select a programme for viewing.

Now the receiver knows which packet identification numbers (PIDs) it's looking for from the PAT, PMT and user input, it can start:

- De-multiplexing the transport packets and recreating the original packetized elementary audio and video streams (PESs). These are then passed to the AV decoder for
- MPEG video and audio decoding (AC-3 decoding for audio in the US), which produces data streams suitable for
- conversion back to analogue base-band audio and video signals. Video outputs will usually be available as RGB or as digitally re-coded PAL, NTSC or SECAM.

All these processes are illustrated in Figure 11.1, which is a block schematic of a typical three-chip digital receiver. Note that, even at this highly schematic level, the actual ICs within a current set-top box already combine many functions together due to an astonishingly high level of integration.

Current set-top box design

The following technical description relates to the Techsan ST1000 family of digital set-top box (STB) satellite receivers. These STBs represent state-of-the art designs, and are therefore representative (in general terms) of technology that the technician will encounter in dealing with digital television equipment in the home and the repair shop (see Figure 11.2). I am very much indebted to Design Sphere (www.designsphere.com) of Fareham (Hampshire, UK) for their co-operation in providing the following information.

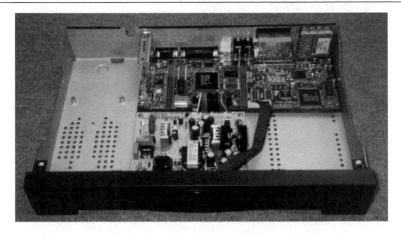

Figure 11.2 *The ST1000 IRD designed by Design Sphere of Fareham in the UK*

Circuit descriptions

The circuit diagrams of this series of the basic ST10000 set-top box are included as Figures 11.3 to 11.8. Each section of the circuitry is described below.

Front end

The front-end (Figure 11.3) consists of the tuner (Tun1) and QPSK demodulator (IC23). The tuner has an input frequency range of 950–2150 MHz. To reduce the effect of noise from other parts of the circuit, a linear regulator (IC26) powers the tuner, providing a clean +5 V supply from the +12 V rail. The output from the tuner is a QPSK (quadrature phase shift keyed) signal. The four-state I and Q outputs are AC coupled into the inputs of IC23, which samples the signals using high-speed A/D converters and then performs the various levels of decoding, de-interleaving and error correction to produce an 8-bit parallel transport data output (labelled TDATA0 ... 7).

IC23 contains a phase locked loop, which is used to multiply up the 15-MHz clock (provided by X3) to derive the sampling clock for the A/D converters. This sampling clock can be up to 90 MHz, depending on the input symbol rate; it can be measured with a high-speed 'scope on the PCLK pin. The filter components for the PLL are C177, R187 and C160. IC23 is split into analogue and digital sections. To reduce noise, the 3.3 V power plane is split underneath the chip. A quiet supply for the analogue section is supplied by L24, C185–186, C194–195. The PLL also has a

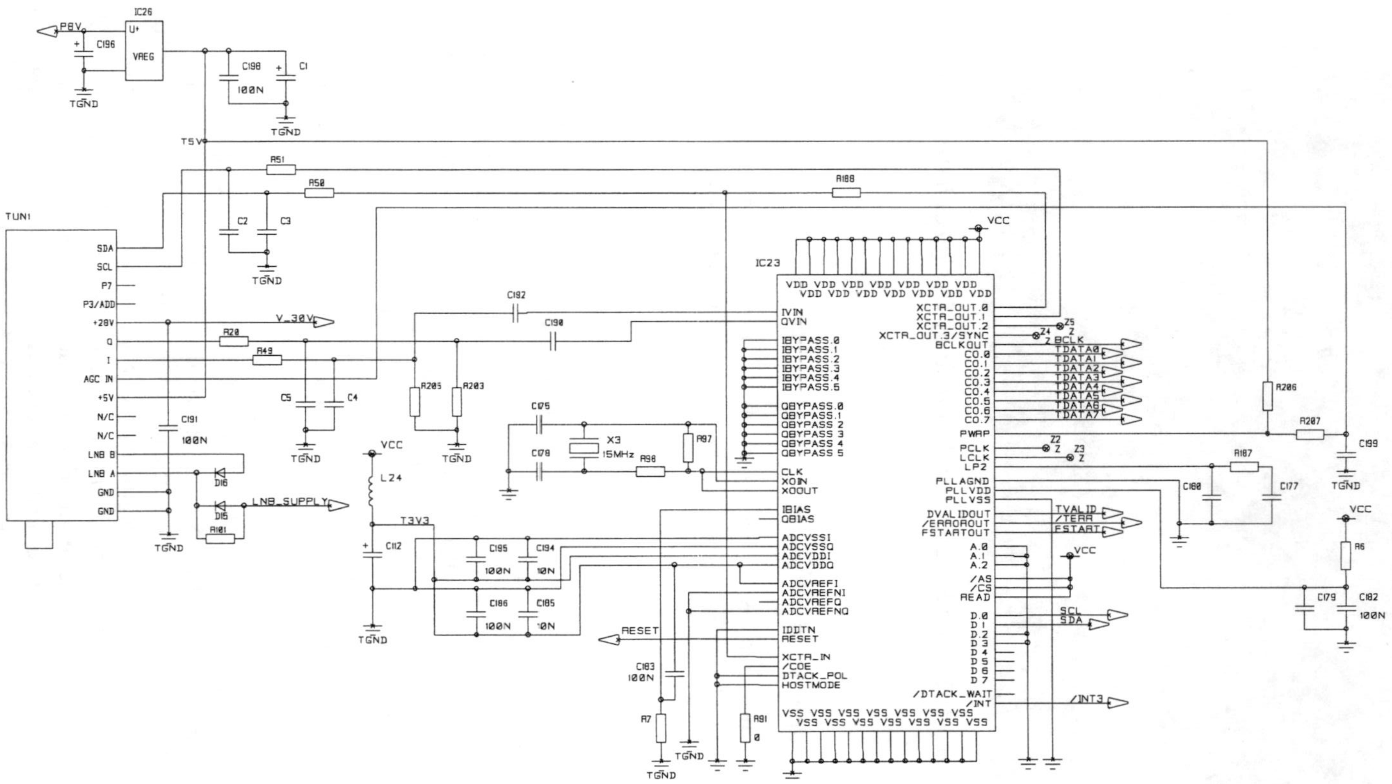

TUN1
SDA
SCL
P7
P3/ADD
+28V
Q
I
AGC IN
+5V
N/C
N/C
LNB B
LNB A
GND
GND
IC26
VREG
P8V
T5V
V_30V
LNB_SUPPLY
VCC
L24
T3V3
X3
15MHz
RESET
IC23
XCTR_OUT.0
XCTR_OUT.1
XCTR_OUT.2
XCTR_OUT.3/SYNC
BCLKOUT
PWRP
PCLK
LCLK
LP2
PLLAGND
PLLVDD
PLLVSS
DVALIDOUT
/ERROROUT
FSTARTOUT
/DTACK_WAIT
/INT
IVIN
QVIN
CLK
XOIN
XOOUT
IBIAS
QBIAS
ADCVSSI
ADCVSSQ
ADCVDDI
ADCVDDQ
ADCVREFI
ADCVREFNI
ADCVREFQ
ADCVREFNQ
IDDTN
XCTR_IN
/COE
DTACK_POL
HOSTMODE
TVALID
/TERR
FSTART
SCL
SDA
/INT3
TGND

Figure 11.3

separate low-noise supply, provided by R6, C179, and C182. To further reduce noise, the tuner is controlled by a separate I²C serial control bus. This only carries data for the tuner, not for the other ICs on the board (they have another I²C bus). The tuner I²C bus is generated by IC23. The I²C signals are also filtered by C2–3, R50–51 to further remove high-frequency noise.

The front-end chip (IC23) also contains AGC circuitry, which is used to control the gain of the tuner and thereby keep the size of the I and Q signals correct (approximately 200 mV pk–pk). The AGC output of IC23 (PWPR) is a digital PWM (pulse-width modulated) signal. The output is open-drain, and R206 is a pull-up to +5 V to give as greater voltage swing on the signal as possible. The PWM signal is filtered by R207 and C199 to produce a DC, AGC voltage, which varies with the mark/space of the PWM output. This signal is fed to the AGC input of the tuner. The front end is controlled via the main I²C bus (SCL, SDA); this is used to set the registers within IC23, and this in turn controls the tuner.

The supply for the LNB (low-noise block) in the satellite dish is passed up the satellite feed cable from the tuner input. This supply is provided by the ST-1000 power supply (see below). The LNB supply can also be provided by another unit connected to the loop out socket of the tuner (from an analogue receiver, for example). This accounts for the inclusion of diodes D15–16, which are used to accommodate the two possible LNB supplies.

The parallel transport data output from IC23 (TDATA0 ... 7) is clocked by the BCLK output. The TVALID and /TERR signals indicate whether the data has errors, and FSTART indicates the start of transport data frames. Turning now to Figure 11.4, you can see that the transport data (TDATA0 ... 7) is available at the expansion connector (J10). A set of 75R resistors (R40, 43, 47, 68, 73, 78, 80, 85, 90, 95, 96) can be removed to intercept the transport data so that it can be passed through the add-on CI option for the ST2000CI common interface STB, which is able to decode scrambled transmissions.

Transport demux

The parallel transport data (now labelled CDATA0 ... 7) is passed to the transport demux IC (IC14, see Figure 11.5). This chip also contains the main MIPS processor, which controls the ST-1000. The software, which runs on IC14, is held in flash memory (IC11). This is directly connected to IC14 by address and 16-bit data buses. IC10 is another flash device, which can be used for memory expansion. DRAM for IC14 is provided by IC19; this is used for transport data buffers and software working area. This is a 4-Mbit device. A footprint for a 16-Mbit device is also provided (IC21) for expansion. Both footprints share the same select signals, so only one

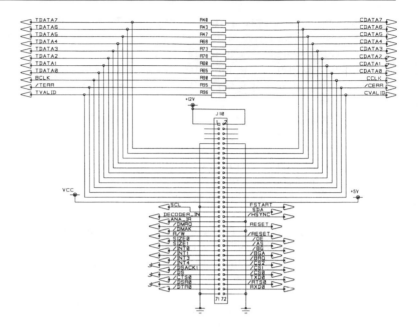

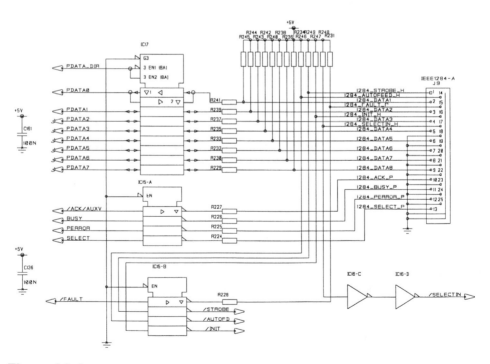

Figure 11.4

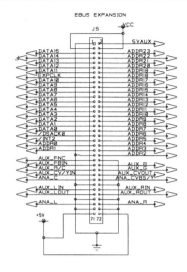

EBUS EXPANSION

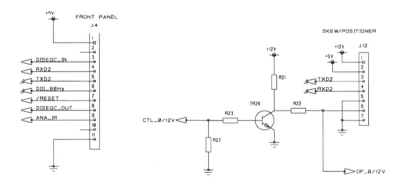

FRONT PANEL
J4

SKEW/POSITIONER

IC16-8

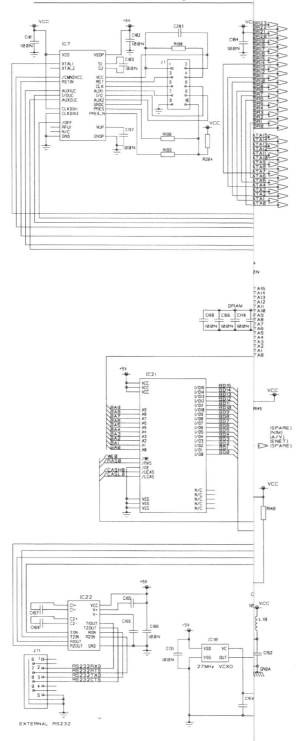

Figure 11.5

device is ever fitted. Refresh of the DRAM is performed by IC14. An EEPROM is provided on the board (IC12) for storage of channel data. This is a 128-kbit I^{2}C device. Alongside IC12 are the pull-up resistors for the I^{2}C bus (R219, 220) – all devices on the bus have open collector outputs. C111 protects against noise pickup on the SCL (I2C clock) signal.

The clock source for IC14 is the 27-MHz VCXO (voltage-controlled crystal oscillator), IC18. IC14 varies the clock frequency to synchronize it with the incoming transport data stream. This is achieved in the same way as we saw with the AGC of the tuner module. In this case, a PWM signal (output SDET) is filtered by R67, C154, 155, 164 to produce a DC voltage which varies with the mark/space of SDET, this filtered signal being fed to the frequency adjust input of the VCXO.

IC14 also has an on-chip 54-MHz clock, which is produced from the 27 MHz input by a frequency doubling PLL. C144, 145, and R77 are the loop-filter components for this PLL. The PLL also has a separate quiet supply, provided by L17, C142, R72, C140 and C201. The 54-MHz clock is present on the MCLK pin of IC14. Yet another clock is generated inside IC14, this being derived from a numerically-controlled oscillator to drive the sample clock (ACLK) of the audio DAC. The frequency of this clock is controlled by software depending on the audio sample rate. The ACLK generator needs a quiet supply, and this is provided by L18, C152, R69, C202, C150. It also requires a reference current, and this is provided by R84 and filtered by C149.

Three serial ports are managed by IC14. Port 0 is connected to the expansion socket (J10); this is reserved for use by expansion options such as a modem and is also used as a debug port during software development. Port 1 passes through an RS232 driver/receiver (IC22) to the rear panel serial port (J11). Port 2 is used to communicate with the ST-1000 front panel (and also with the add-on dish skew positioner, connected via J12). An IEEE1284 port (enhanced printer port) is provided on the rear panel (J9 – see Figure 11.4). This is controlled directly by IC14 under software control, and provides a point at which to pick-up parallel transport data should this be required in another product. The high-speed parallel signals are buffered by IC17 and IC15 (and 1/3 of IC16).

The transport demux chip (IC14) processes the incoming transport stream under software control and filters out the required MPEG encoded data for the bouquet of TV and/or radio channels available on the multiplex (a process that is usually referred to as PID filtering). The PID filtered data is sent to the A/V decode chip (IC4 – see Figure 11.6) via the 8-bit parallel A/V data bus (labelled AVD0 ... 7). Signals /AREQ and /VREQ are handshake signals, which IC4 uses to request audio and video data from IC14. In other words, it is the MPEG decoder block (A/V decoder) that regulates the flow of input information it requires. Sufficient data must always be available to avoid sound or picture break-up.

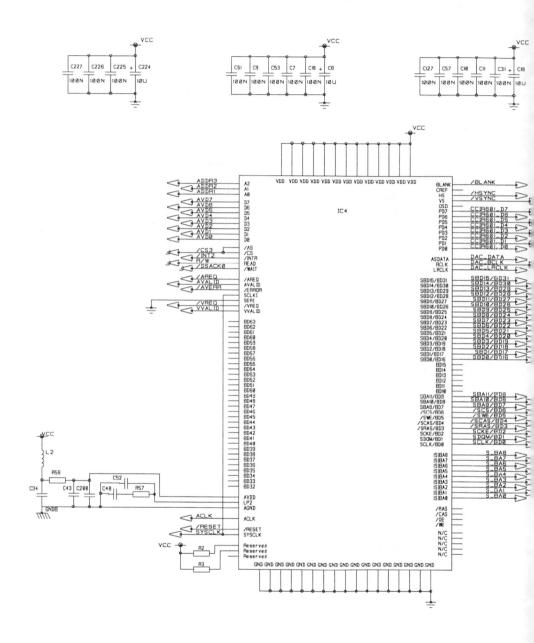

Figure 11.6

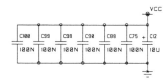

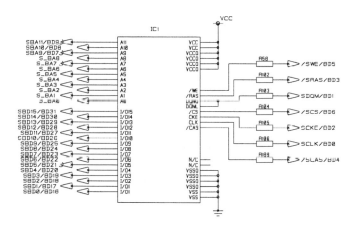

AVALID and VVALID signals indicate that the data being sent on AVD0 . . . 7 is valid.

Teletext™

Analogue Teletext was explained in Chapter 2. If a selected digital TV channel contains Teletext information, it is IC14 that is responsible for 'filtering out' this information from the transport stream data. But how is this re-injected into the video output so that it may be decoded in the conventional way? Looking at the schematic, you can see that Teletext information is passed from IC14 direct to the PAL/NTSC encoder (IC6 in Figure 11.6) to be included in the video waveform to the TV (signal TTXDATA). TTXREQ is the handshake signal from IC6 to synchronize the data with the correct lines of the video frame.

A/V decoder

The A/V decoder (IC4) takes the data from the transport demux IC and performs MPEG decompression to provide base-band, audio and video data for the PAL/NTSC encoder and audio DAC. It also contains the graphics hardware needed to produce OSD (on-screen display). To provide the necessary memory to buffer the audio, video and OSD data, IC4 has memory attached to it. IC4 uses the 27-MHz clock as its source clock. It has an on-chip PLL to generate an 81-MHz clock, which it uses to control (S)DRAM access timings. The PLL has a separate quiet supply, provided by L2, C34, R56, C43 and C200. The filter components for the PLL are included on-chip.

The video data to the PAL/NTSC encoder is output as an 8-bit, parallel data bus (CCIR601_D0 . . . 7). This data must be synchronized to the timings of the output video waveform. This is achieved by the (/HSYNC) and (/VSYNC) signals. IC4 also outputs a video blanking signal (/BLANK). The audio data to the audio DAC (IC2) is output as DAC_BCLK (bit clock), DAC_DATA, and DAC_LRCLK (left/right clock) signals. IC4 uses ACLK (audio sample clock), generated by IC14, to synchronize the audio data.

PAL/NTSC encoder

The PAL/NTSC encoder (Figure 11.7) converts the digital data from the A/V decoder into video output to be routed to a standard TV or VCR. The supply for IC6 is split into digital and analogue sections. L5, C63 and C56 isolate the analogue supply. On-chip references are decoupled by C67, 68 and C69. The full-scale output voltage of the on-chip D-to-A converters, used to generate the video waveform, is set by R61. IC6 has four outputs; these are designed to drive a double terminated 75R load. The outputs are green, blue and red/Chroma and composite video/Luma, the latter two depending on whether RGB or S-Video modes are selected. Each output passes through a filter stage to remove D/A conversion artefacts.

Audio DAC

The audio DAC (Figure 11.8) takes audio data from the A/V decoder and converts it to analogue audio. The data are synchronized to the audio sample rate by the ACLK signal from the transport demux. This signal is 384 times the audio sample rate. The stereo audio output of the DAC is filtered by IC3 to remove the out-of-band noise caused by the D/A conversion.

Power supply unit (PSU)

The power supply fitted to the ST1000 is a bought-in item, designed specifically for use with these products. There are several manufacturers of this supply, but all of the supplies have the same rating, mountings and input/output connectors; they are therefore completely interchangeable. (The PSU board can be seen clearly in Figure 11.2.) All the supplies are off-line switched-mode type, and are each based on a combined control/power switching IC. The PSU in a digital set-top box often has to provide a considerable number of different voltage rails. The front-end, transport demux and AV decode 3-chip sets use mostly 3.3 V logic (some manufacturers are now talking about 1.8V!). PAL encoder chips and glue logic and CI board still typically require 5 V supplies. Varicap tuning will usually be accomplished with a *circa* 30 V supply, and the analogue output circuits and the tuner module may require +12 V (and sometimes −12 V). The LMB will also require a supply in the case of a satellite transmitter (usually in the range of +12 V to +24 V). The nominal tolerance on all supply rails is ±5 per cent. The supplies are universal input types; the input supply range is 90–250 V AC without need for range setting or adjustment.

Set-top box – modern trends

Digital tuner

The tuner module is still an expensive item (even when manufactured in the Far East in vast quantities), and in this highly price-sensitive market every penny counts. One can therefore predict a move towards a digitization of the tuning functions; so that RF will be directly converted to digital PCM for mixing and IF filtering. This technique is sometimes referred to as 'zero IF'.

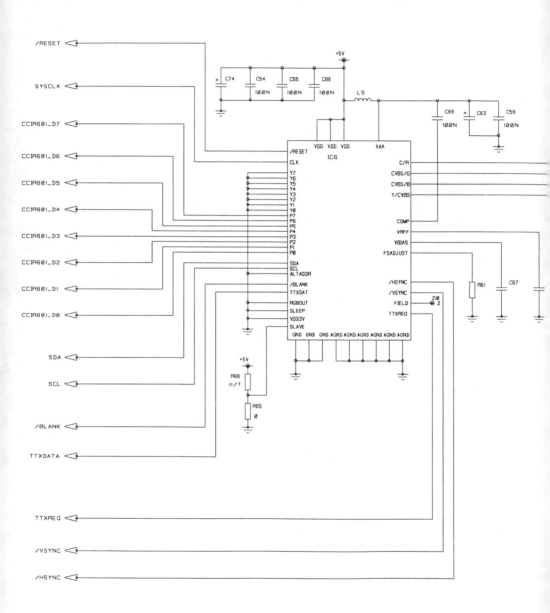

Figure 11.7

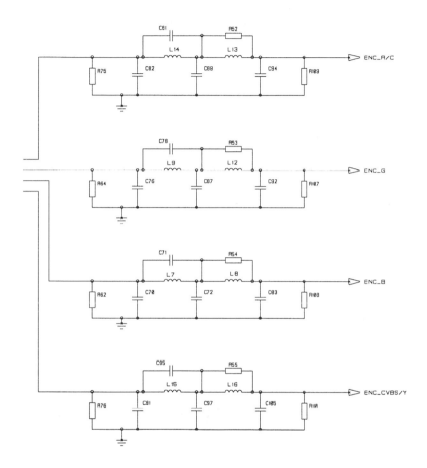

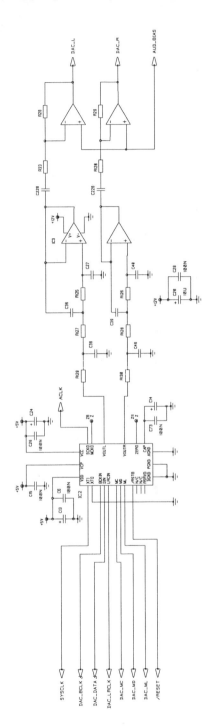

Figure 11.8

Incorporation of hard disk drives (PVR)

The signal that appears after the transport demux chip is in a form suitable for recording onto a hard disk drive. Why should you want to do this? Because it would provide the ability to record programmes for time-shift purposes. Customer research has shown this is the main function of video recorders; not the long-term archiving of material, but rather the shifting of certain programmes to be watched at convenient times. A disk drive enhances the power of the set-top box and moves it effectively into being a short-term video recorder, nowadays known as the PVR or the Personal Video Recorder.

COFDM front-end for DTV-T

COFDM presents some very special front-end problems – the reality of performing a very high speed 8000 (or 2000) FFT being the main issue! A chip from Fujitsu Microelectronics demonstrates the complexity of the front-end operations of a modern set-top box for terrestrial transmission. The MB87J2050 is illustrated in Figure 11.9. First, the incoming OFDM signal is shifted in the frequency domain in order to have it centered around the zero axis. Data is processed in an 8000 or 2000 FFT, which recovers the data on each individual carrier. The resulting data is de-interleaved and passed to the de-mapper before passing to the Viterbi decoder and Reed-Solomon decoder. Data is output on the CI, as shown.

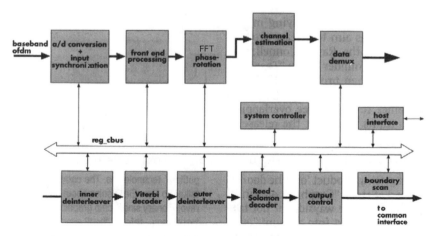

Figure 11.9 *Fujitsu front-end chip for DVB-T*

D-VHS

D-VHS is Japanese electronic giant JVC's continuation of the ubiquitous analogue VHS format. (Incidentally, the 'D' stands for data, not digital.) D-VHS shares many common components with a conventional VHS player, and accommodates analogue recording of analogue sources. Table 11.1 annotates the general specification of a current D-VHS machine.

D-VHS tape is an improved VHS type in an identical shell except for two extra holes for identification. Digital input and output is provided on an IEEE 1394 interface (Firewire), which we have already met in Chapter 8. IEEE 1394, which was originally Apple Computers Firewire network system, allows transfer speeds of up to 400 Mbits/s to 63 nodes on the network. Firewire supports both asynchronous transfers, which amount to usual data-type traffic over a computer network, and isochronous transfers, which are reserved for time-critical data – such as video and audio signals which need to be real-time to avoid frame freezes. The proportion of the connection allocated for isochronous data packets (yes, this too is another packetized interface!) is 'negotiated' during each network session by the root node, the latter being defined during the initialization phase.

In the case of recording digitally on the proposed D-VHS, the set-top box would need to output MPEG-II multiplex; however, most boxes (as described above) do not provide an interface at this point in the data stream. Future boxes will need to incorporate this interface for full

Table 11.1   *General specification of a current D-VHS machine*

Cassette	188 × 104 × 25 mm
Tape	D-VHS 0.5 inch oxide
Tape speed	16.67 mm/s
Drum diameter	62 mm
Drum rotation	1800 rpm
Tracking	CTL track
Recording	Azimuth ±30 degrees
Track pitch	29 μm
Channel coding	SI-NRZI
Error correction	Reed-Solomon 6-track interleaved
Tape data rate	19.1 Mbits/s
Nett data rate	14.1 Mbits/s (peaks to 72 Mbits/s)
Digital interface	IEEE 1394, Apple 'FireWire'
Input signal	MPEG-II transport stream
Output signal	MPEG-II transport stream

integration with a digital record facility. The MPEG-II transport output of the set-top box would be the root node and would send MPEG-II transport stream to the digital input of the D-VHS recorder. Note that, at an overall data record rate of 14 Mbits/s, the recorder can theoretically record all the current programmes on the multiplex. Note also that replay may require descrambling, and in this way piracy can be prevented because recordings could have time-stamps which allow decoding only for a period of time.

Physically, the IEEE 1394 interface uses a four or six-pin connector which carries two (or optionally three) pairs of connections. The first two pairs; DATA and STROBE, are both transmitted differentially, and the third pair may be present for power. The interface is terminated at both ends, and can only run 4.5 metres. The power pair is deemed useful because, in a consumer environment, the network would be 'broken' if one node on the network was switched off. Each node can therefore be powered externally by the power connections arriving in the interface. D-VHS may yet threaten the new DVD format discussed below due to its backward compatibility, with the viewer being able to play original analogue tapes on the same machine.

DVD

Unlike a CD, a DVD (digital versatile disc) has the capacity to be double-sided. Two thin (0.6 mm) back-to-back substrates are formed into a single disc that's the same thickness (1.2 mm) as a regular CD but more rigid. Data is represented on a DVD as it is on a CD: by means of physical 'pits' on the disc. But the thinner DVD substrates (and short-wavelength visible light laser) permit the pits to be smaller. In fact, they're roughly half the size, which in turn allows them to be placed closer together. The net effect is that DVDs have the capacity for over four times as many pits per square inch as CDs, totalling some 4.7 Gb in a single-sided, single-layer disc.

A DVD's capacity may be further increased by employing more than one physical layer each side of the disc! In this case, the inner layer reflects light from the laser back to a detector through a focusing lens and beam-splitter because the outer layer is only partially reflective. DVD players incorporate novel dual-focus lenses to support two-layer operation, yielding 8.5 Gb in a single-sided DVD, or 17 Gb in a double-sided disc.

Track structure

There are three types of track structure, depending on the type of disc. For a single layer disc, the track is formed as continuous spiral from inside to outside of disc (just as in a conventional CD), like this:

```
BBBBBBBBBBBBBBBBBBBBBBBBBBBBBBBB    outer edge
XXIIIDDDDDDDDDDDDDDDDDDDDDDDOOOXX    of disc
```

where the letters represent the following areas:

I   lead-in area (leader space near edge of disc)
D   data area (contains actual data)
O   lead-out area (leader space near edge of disc)
X   unusable area (edge or hole)
M   middle area (interlayer lead-in/out)
B   dummy bonded layer (to make disc 1.2 mm thick instead of 0.6 mm)

For a dual layer disc as in DVD-ROM used in a computer, the track direction is the same for both layers.

```
XXIIIDDDDDDDDDDDDDDDDDDDDDDDOOOXX   Layer 1
XXIIIDDDDDDDDDDDDDDDDDDDDDDDOOOXX   Layer 0
```

For a dual layer DVD-video disc for a continuous movie film, the track direction is a spiral on layer 0 from the centre of the disc to the edge, transitioning to a spiral track from the edge of the disc to the centre on layer 1. This is to permit a seamless (or nearly seamless) transition as the player switches between layers and can be thought of a bit like an auto-reverse cassette deck!

```
XXOOOODDDDDDDDDDDDDDDDDDDDMMMXX   Layer 1
XXIIDDDDDDDDDDDDDDDDDDDDDDDMMMXX   Layer 0
```

Data rates and picture formats

DVD-video is an application of DVD-ROM. DVD-video encodes video pictures according to the MPEG-II standard and – at a rough average rate of 4.7 Mbit/s (3.5 Mbit/s for video, 1.2 Mbit/s for three 5.1-channel soundtracks) – a single-layer DVD can hold a little over two hours. A dual-layer disc can hold a two-hour movie at an average of 9.5 Mbit/s (close to the 10.08 Mbit/s limit). A disc has one track (stream) of MPEG-II constant bit rate (CBR) or variable bit rate (VBR) compressed digital video. A restricted version of MPEG-II Main Profile at Main Level (MP@ML) is used. SP@ML is also supported as is MPEG-I CBR and VBR. 525/60 (NTSC, 29.97 interlaced frames/s) and 625/50 (PAL, 25 interlaced frames/s) video display systems are expressly supported. Note that very few players do standards conversion: it simply follows the MPEG-II encoder's instructions to produce the predetermined display rate of 25 fps or 29.97 fps.

Allowable picture resolutions are:

MPEG-II, 525/60 (NTSC): 720 × 480, 704 × 480, 352 × 480
MPEG-II, 625/50 (PAL): 720 × 576, 704 × 576, 352 × 576
MPEG-I, 525/60 (NTSC): 352 × 240
MPEG-I, 625/50 (PAL): 352 × 288

Maximum video bit rate is 9.8 Mbit/s, which is substantially better than most transmitted digital TV. Typical video bit rate is 3.5 Mbit/s (nearer typical TV transmission rates) but this depends on the length, quality and amount of audio. Raw channel data is read off the disc at a constant 26.16 Mbit/s. After 8/16 demodulation the effective rate is halved to 13.08 Mbit/s. (8/16 coding is the channel coding used for DVDs in which 16 recorded data bits translate to eight useable data bits. The 8/16 channel code helps reduce DC energy – runs of '1' or '0' bits – thereby lowering the SNR threshold for the pickup signal.)

After error correction the user data stream goes into the track buffer at a constant 11.08 Mbit/s. The track buffer feeds system stream data out at a variable rate of up to 10.08 Mbit/s. After system overhead, the maximum rate of combined elementary streams (audio + video + subpicture) is 10.08 Mbit/s. Still frames (encoded as MPEG-II, I-frames) are supported. These are used for menus and can be accompanied by audio. A disc also can have up to 32 subpicture streams that overlay the video. These are used for subtitles and captions for the hard of hearing. These sub-pictures can be full-screen, run-length-encoded bitmaps. The maximum subpicture data rate is 3.36 Mbit/s.

DVD supports 4:3 and 16:9 aspect-ratio video. DVD players have four playback modes, one for 4:3 video and three for 16:9 video:

• full frame (4:3 video for 4:3 display)
• auto letterbox (16:9 anamorphic video for 4:3 display)
• auto pan and scan (16:9 anamorphic video for 4:3 display)
• widescreen (16:9 anamorphic video for 16:9 display)

Audio

A DVD-video disc can have up to 8 audio tracks (streams) associated with a video track. Each audio track can be in one of three formats:

• Dolby Digital (AC-3): 1 to 5.1 channels
• MPEG-II audio: 1 to 5.1 or 7.1 channels
• PCM: 1 to 8 channels.

Two additional optional formats are provided: DTS and SDDS. Both require external decoders and are not supported by all players. Linear PCM can be sampled at 48 or 96 kHz with 16, 20, or 24 bits/sample. Discs containing 525/60 video (NTSC) must use PCM or Dolby Digital on at least one track. Discs containing 625/50 video (PAL/SECAM) must use PCM or MPEG audio or Dolby Digital on at least one track. Additional tracks may be in any format. For stereo output (analogue or digital), all players have a built-in 2-channel Dolby Digital decoder that downmixes from 5.1 channels (if present on the disc) to Dolby Surround stereo (i.e., 5 channels are phase matrixed into 2 channels to be decoded to 4 by an external Dolby Pro Logic processor). PAL players also have an MPEG or MPEG-II decoder. Both Dolby Digital and MPEG-2 support 2-channel Dolby Surround.

Control

DVD-video players support a command set that provides rudimentary interactivity. The main feature is menus, which are present on almost all discs to allow content selection and feature control. Each menu has a still-frame graphic and up to 36 highlightable, rectangular 'buttons'. Remote control units have four arrow keys for selecting on-screen buttons, plus numeric keys, select key, menu key, and return key. Additional remote functions may include freeze, step, slow, fast, scan, next, previous, audio select, subtitle select, camera angle select, play mode select, search to program, search to part of title (chapter), search to time, and search to camera angle. The producer of the disc can disable any of these features.

DVD-video content is broken into 'titles' (movies or albums), and 'parts of titles' (chapters or songs). Titles are made up of 'cells' linked together by one or more 'program chains' (PGC). A PGC can be one of three types: sequential play, random play (may repeat), or shuffle play (random order but no repeats). Individual cells may be used by more than one PGC, to allow parental management and seamless branching.

Regional codes

DVD technology is generally used for movies and the motion picture studios want to control the home viewing of movies in different countries because of different rights arrangements. To that end, there are six regional codes that apply to DVD which are illustrated Figure 11.10. Generally the regional code will be located on the back bottom of the DVD case. The regions are:

1.   Canada, the USA, US Territories
2.   Japan, Europe, South Africa, Middle East (including Egypt)

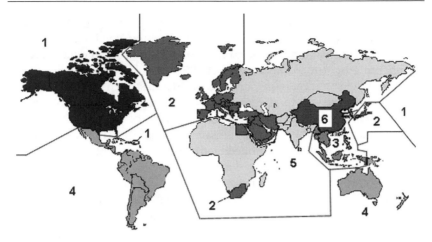

Figure 11.10 *DVD geographical regions*

3. Southeast Asia, East Asia (including Hong Kong)
4. Australia, New Zealand, Pacific Islands, Central America, Mexico, South America, Caribbean
5. Former Soviet Union, Indian Subcontinent, Africa (also North Korea, Mongolia)
6. China.

DVDs having an inappropriate region code will not play and the player will usually display the message, 'Illegal Region Code!'.

The DVD player

Obviously, due to the MPEG coding, the final stages of a DVD player resemble those of the set-top box described above. However, there are clearly a number of stages that precede this to do with the reading and decoding of the data from the physical disc. Prior to the decoder, the flow of data in a DVD player may be broken down into a number of steps. The DVD data stream is encoded using 8/16 modulation, and the first stage within the player is the reading and demodulation of this code. But, for this to be achieved the read must be synchronized. This is achieved by means of special sync words inserted in the modulation code. Sync code words are unique in the 8/16 code table (so they cannot be generated by the 8-to-16 mapping). The detection stage looks for sync codes in order to determine where sectors begin and end. At this point the 26.16 Mbit/s stream from the disc is reduced to 13.08 Mbit/s.

This initial stage is followed by error detection and correction. In a similar way as we saw with broadcast TV, DVD protects the data by the addition of error (Reed-Solomon) coding. If the check bits (EDC) don't match the fingerprint of the unscrambled data, the Reed-Solomon bytes (IEC) are used to attempt error correction of the corrupted data. Here the channel rate output by this block is 11.08 Mbit/s because approximately 2 Mbit/s of error correction parity data, IEC, has been stripped. Data is subsequently passed to the track buffer. This FIFO (first in first out buffer) maps the constant user data bit rate of 11.08 Mbit/s to the variable bit rate of the program streams. DSI and PCI packets (used to control the behaviour of the player) are stripped yielding a 10.08 Mbit/s rate into the MPEG systems decoder. The mux_rate of all program streams is 10.08 Mbit/s regardless of actual elementary stream rates. The size of the track buffer is left to the implementation, although the minimum recommended size is 2 Mbit. Finally, data is transferred to MPEG system decoder.

CPSA (content protection system architecture)

CPSA is the name given to the overall framework for security and access control across the entire DVD family. There are many forms of content protection that apply to DVD. All these copy protection schemes are designed to guard against casual copying only: none of them will stop well-equipped pirates.

Analogue copy protection

Analogue copying is prevented with a 'Macrovision', 'Copyguard' or a similar circuit in every DVD player. These systems add a rapidly modulated colour-burst signal ('Colorstripe') and pulses in the vertical blanking. These extra signals confuse the synchronization and automatic-recording-level circuitry in 95 per cent of consumer VCRs. These protection systems were not present on the analogue component video output of early players, but are required on newer players. Just as with videotapes, some DVDs are protected in this way and some aren't, depending on whether the producer has opted to pay the appropriate licence.

Copy generation management system (CGMS)

Each disc also contains information specifying if the contents can be copied. The CGMS information is embedded in the outgoing video signal. For CGMS to work, the equipment making the copy must recognize and respect the CGMS information. The analogue standard (CGMS-A) encodes the data on NTSC line 21 or line 20. CGMS-A is recognized by most digital camcorders and by some computer video capture cards (they will flash a message such as 'recording inhibited'). Professional time-base correctors (TBCs) that regenerate lines 20 and 21 will remove CGMS-A information from an analogue signal. The digital standard (CGMS-D) is not yet finalized.

Both these systems are pretty useless because it's relatively easy to remove the Macrovision signals and to remove and replace lines 20 and 21.

Content scrambling system

Content scrambling system (CSS) was an attempt at a more complete answer to data encryption and authentication. CSS was intended to prevent copying video files (.vob format) directly from DVD-video disks. Each CSS licensee is given a key from a master set of 400 keys that are stored on every CSS-encrypted disc. The CSS decryption algorithm exchanges authentication and decryption keys prior to disc playback.

Unfortunately CSS has now been cracked and there exist several software utilities, freely available on the Internet. The code was cracked by a group of Norwegian software hackers who discovered one of the 400-odd companies licensed to use CSS had failed to encrypt the unique software 'key', which is used to unlock the scrambled content on the disc. Having cracked one key, it was only a matter of days before the other encryption keys applicable to all DVD movies and players were decoded. This means that anyone with a DVD-ROM fitted to their PC can transfer movies directly to their hard drives. With recordable DVD-RAM systems now available, perfect DVD movie copies are now a reality.

Other systems of copy protection exist: Content Protection for Pre-recorded Media (CPPM) is used only for DVD-audio. It will not be discussed here. Content Protection for Recordable Media (CPRM) is a mechanism that ties a recording to the media on which it is recorded. It is supported by all DVD recorders released after 1999. Each blank record-able DVD has a unique 64-bit media ID. When protected content is recorded onto the disc, it can be encrypted with a 56-bit C2 (Cryptomeria) cipher derived from the media ID. During playback, the ID is read from the disc and used to generate a key to decrypt the contents of the disc. If the contents of the disc are copied to other media, the ID will be absent or wrong and the data will not be decipherable. Some systems focus on protection via the digital interface, to prevent perfect copies. DTCP (Digital Transmission Content Protection) focuses on IEEE 1394/FireWire. Sony released a DTCP chip in mid 1999. Under DTCP, devices that are digitally connected, such as a DVD player and a digital TV or a digital VCR, exchange keys and authentication certificates to establish a secure channel.

DVD Recordable (DVD-R)

Similar in concept to CD-R, DVD-R is a write-once medium that can contain any type of information normally stored on mass-produced DVD discs. Depending on the type of information recorded, DVD-R discs are usable on any DVD playback device, including DVD-ROM drives and DVD video players. A total of approximately 7.9 or 9.4 Gb can be stored on a two-sided DVD-R disc. Data can be written to or read from a disc at 11.08 megabits per second (Mbit/s), which is roughly equivalent to nine times the transfer rate of CD-ROMs '1X' speed. DVD-R, like CD-R, uses a constant linear velocity rotation technique to maximize the storage density on the disc surface. This results in a variable number of revolutions per minute (RPM) as disc writing/reading progresses from one end to the other. To achieve a sixfold increase in storage density over CD-R, two key components of the writing hardware needed to be altered: the wavelength of the recording laser and the numerical aperture (n.a.) of the lens that focuses it. In the case of CD-R, an infrared laser with a wavelength of 780 nanometers (nm) is employed, while DVD-R uses a red laser with a wavelength of 635 nm. These factors allow DVD-R discs to record 'pits' as small as $0.44\,\mu m$ as compared with the minimum $0.834\,\mu m$ size with CD-R.

Recording on DVD-R discs is accomplished through the use of a dye polymer recording layer that is permanently transformed by a highly focused laser beam. This dye polymer substance is spin-coated onto a clear polycarbonate substrate that forms one side of the 'body' of a complete disc. The substrate has a microscopic, 'pre-groove' spiral track formed onto its surface. This groove is used by a DVD-R drive to guide the recording laser beam during the writing process. A thin layer of metal is then sputtered onto the recording layer so that a reading laser can be reflected off the disc during playback. The recording action takes place by momentarily exposing the recording layer to a high power (10 mW) laser beam that is tightly focused onto its surface. As the dye polymer layer is heated, it is permanently altered such that microscopic marks are formed in the pre-groove. These recorded marks differ in length depending on how long the write laser is turned on and off, which is how information is stored on the disc. The light sensitivity of the recording layer has been tuned to an appropriate wavelength of light so that exposure to ambient light or playback lasers will not damage a recording. Playback occurs by focusing a lower power laser of the same approximate wavelength (635 or 650 nm) onto the surface of the disc. The 'land' areas between marks are reflective, meaning that most of the light is returned to the player's optical head, whereas, recorded marks are not very reflective, meaning that very little of the light is returned. This 'on-off' pattern is thereby interpreted as the modulated signal, which is then decoded into the original user data by the playback device.

General servicing issues

Static and safety

Remember that VLSI integrated circuits are sensitive to damage by static. A wise service technician will always use an anti-static wrist strap and properly grounded anti-static work surface when touching or removing modern internal sub-assemblies or components. Removed sub-assemblies must also be properly packed in anti-static packing material. Remember too that PSUs often remain powered, even when in standby mode. Therefore, always disconnect equipment from the mains supply when dismantling or re-assembling. When testing with the lid removed it is strongly recommended that the unit be powered from a mains isolating transformer, as this will minimize the risk of electric shock due to accidental contact with the power supply. Failing this, a home-made PSU cover made from stiff card and taped to the chassis will prevent many uncomfortable shocks.

Golden rules

Modern integrated circuits are amazingly reliable. Despite the complexity of modern domestic digital television equipment, most faults experienced will be of mechanical origin; either outside the house (aerial and cable problems for digital terrestrial, problems with antenna, head-unit support and the feedhorn for digital satellite equipment) or inside the house (innocently re-arranged wiring, mis-plugging of peripheral equipment etc.). In the case of D-VHS and DVD, where a mechanism is part of the machine, it is nearly always the physical mechanism that gives trouble rather than the electronics. When faced with diagnosing a fault, always be sure to check the obvious before proceeding on the assumption of a more complex hypothesis. In addition, always remember that a customer will sometimes, when faced with a problem, make things worse by adjusting tuning controls etc. So a simple plugging or mechanical fault may, by the time you arrive, be compounded by several layers of mis-adjustment. The golden rule is therefore to make sure that your analysis of a fault is always logical and systematic, and remember that clients will often, out of chagrin, deny having adjusted the controls and thereby made things worse! Finally, remember that digital television 'stops' at the SCART connector. TV line and field timebases, PSUs and video amplifiers remain medium-power linear circuits subject to stress and strenuous design cost constraints.

Equipment

The choice of equipment for the service engineer in the digital age is difficult. Frankly, commercial considerations will probably win-out over purely technical ones. The commercial conundrum for the modern service engineer is this; a modern digital set-top box is a fiendishly complicated piece of equipment. It is difficult, if not impossible, to find detailed service data for many models, and the money is made by selling the 'package' of set-top box and programme services rather than the equipment itself. Thus a box, which might take many hours to fault-find, will have been sold for the money equivalent of a few hours service work, thereby invalidating the fiscal advantage of repair over re-purchase. There is a parallel here with mobile phones (indeed, a modern set-top box owes a great deal technologically to the portable communications revolution), where the telephones themselves are literally given away and it is the air-time contract that is sold; a situation which, coupled with the inaccessibility of service information and the sophistication of modern electronics, hardly encourages the repair industry!

Nevertheless, as already stated, modern ICs are very reliable, and therefore most faults (and certainly those faults which will be economic to repair) will be of a very simple nature and will require only the most basic service equipment. In fact, in this respect at least the digital revolution is no cause for concern, and many faults can be found with nothing more that a multimeter and a signal strength meter (certainly for satellite work). An oscilloscope is an advantage as well, although for signal work you will need an instrument with a 250-MHz response at least. Most faults will also be repairable using nothing more than a conventional controlled soldering iron and simple tools.

DVD faults

DVD players represent one of the only real hazards to the service engineer, because the laser must never be viewed directly or serious damage to eyesight will result. Very considerable care must be taken when working with DVD in this respect. Service engineers familiar with CD players should find much in common with DVD machines (which are, after all, only advanced CD players). Once again, problems are often as simple as a dirty or damaged disk, or tuning or plugging problems with the TV set itself. The other mechanical area that can give trouble is the automatic tray loading mechanism. This is itself often responsible for damage to disks, causing them to be scratched in the process of loading, or it may fail to open or close due to the faulty action of an interlock micro-switch. Another possible area of trouble is the disk motor which, if it is worn or faulty, will often rotate the disk, but the disk will not play.

Problems with the optical pickup unit can be diagnosed from viewing the RF waveform as in a CD player, provided service information is available. The various servo-controlled mechanisms can cause problems too; both the focus and the tracking servos. However, these are rarely of electronic origin, although they can result from the ageing or oxidization of pre-set adjustments for servo offset and gain. This type of fault can usually be repaired, but special test disks and service information are often necessary to avoid exploratory 'tweaking' – which can in itself result in a player that is rendered useless by careless adjustment of a system parameter that can only be subsequently readjusted by using specialized (and absent) test equipment.

PSU faults

Power-supply faults are the most common electronic faults. The components within the PSU are virtually always the most stressed system components, and are therefore the most likely to fail. Fortunately for the service engineer, the situation is little different to analogue equipment, and very many faults can be diagnosed with little more than a multimeter and eyesight. Of all components within the PSU itself, the most likely to fail are fuses or fusible resistors, reservoir capacitors, linear regulators (often the 78xx and 79xx series) and, in the case of switched-mode power supplies, the switching transistors.

Inspection of Figures 11.3 to 11.6 shows that there are many subsmoothed supplies throughout modern equipment. These are often supplied through small inductors, which are fragile and are quite often the cause of localized PSU faults.

In each case these types of faults are relatively easy to find and repair, although care should always be taken to replace components with those of an equivalent type and voltage rating. Power supply rectifier diodes can often go open circuit too; once again, this is usually fairly easy to establish. Remember that most PSU faults may be reported to you in a fashion that suggests something much more complex. This is because, due to the proliferation of supplies in most digital equipment, many sub-systems will continue to operate; it is only the simplest of faults that result in total system shut down.

Care should be taken when repairing PSUs, especially after replacing fuses. Often a fuse has blown for a good reason, and on re-connection will blow again immediately. Never replace a fuse with a high rating type in order to trace a downstream fault; this can result in the explosion of capacitors and even fire. When reconnecting a PSU after replacing a component, always stand back in case red-hot electrolyte issues from somewhere!

12
The future

Leaning forward and leaning back

The incredible changes wrought in television in the last few years did not happen in a technological and commercial vacuum. In truth, television is busy redefining new roles for itself as it competes against the world of multimedia computing. It may well be that television, as we know it now, will eventually be swallowed-up into a high bandwidth World Wide Web (WWW). This chapter examines some of the forces at work in television today, and predicts (by looking at the current work of the Moving Picture Experts Group) what the future will hold as this century progresses. Business pages of newspapers tend to regard the competition between television and multimedia computing as the great theatre of technological war in the next 10 years. Certainly there will be losers and winners but, for the consumer, the divisions will be much less obvious than the gradual alloying of previously disparate technologies. Nowhere is this clearer than in the living-room approach to television versus browsing the WWW on the Internet. With the cultural explosion of the Internet, television companies everywhere are searching for ways to incorporate some of the 'Net experience' into the television experience. The difference in viewing styles has been termed the 'lean-forward' experience of browsing versus the 'lean-back' experience of traditional television programming.

One obvious way to counteract the haemorrhaging of viewers away from television to their computers is to include WWW (hypertext mark-up language or HTML) pages along with the audio and video information in the MPEG-II multiplex and to provide a browser (and memory) within the set-top box to provide associated information, perhaps at a deeper level, to be browsed along with – or after – the programme. This would allow the TV viewer to migrate from 'leaning back' to 'leaning forward' when they want to.

Hypertext and hypermedia

Hypertext is text that need not be read linearly in the sequence it is initially presented. In contrast with a physical book, hypertext organizes text not as a sequence of pages held by a strong binding, but as a loose web of programmed interconnections. Provided the author or editor has provided links in the information that point to other relevant portions of the text, the reader is free to hop around the text by means of these links rather than read slavishly from beginning through to the end. Hypermedia extends the concept of hypertext so that pictures, media clips etc. can be included within the text.

HTML documents

Hypertext mark-up language (HTML) is the data format used in the World Wide Web to instruct the browser program how to present a particular document. HTML information can easily be sent in the MPEG-II multiplex – at a certain PID – provided that it is identified within the various tables as such and that the set-top box has the technology to parse the files and display them in the way that a WWW browser does. The method of programming the all-important hypertext and hypermedia links is described here.

Every HTML document (file) is divided into two parts; a head and a body. The head contains information about the document, and the body contains the text of the document itself. Usually, the only information that needs to go in the head section is the title of the document. Every HTML document should have a short, relevant title. This will usually be displayed as the title of the window when interpreted by the browser program. The body of the document contains one paragraph of text, although the term paragraph is used in a special way here. The text in the body of the document does not contain word-wrap instructions; this is because the browser program deals with word wrapping. The advantage of this is that different readers can set the window size of their browser programs differently, and the browser will automatically fill the window with text. If and when line breaks and paragraph (used in the normal sense) breaks are required, they must be written in a form the browser can interpret. Borrowing from the parlance of the printer, these mark-up instructions are known as mark-up tags. There are a significant number of mark-up tags, and all of them are written and segregated from ordinary text by means of brackets. Some of the most common are illustrated in this simple HTML document.

```
<HEAD>

<TITLE>Hypertext</TITLE>
```

< /HEAD>

< BODY><H1>Hypertext and Hypermedia</H1>

<EM>Hypertext</EM> is text which need not be read line-
arly, in the sequence it is initially presented. In contrast with a physical book, hypertext organizes text not
as a sequence of pages held by a strong binding, but as a
loose web of programmed interconnections. Provided the
author provides links in the information which point to
other relevant portions of the text, the reader is free to
hop around the text by means of these links, rather than
read slavishly from beginning through to the end.

<P>Hypermedia extends the concept of hypertext so that
pictures, media clips etc. can be included within the
text.</BODY>

Let's look at some of the mark-up tags used. Most of them are self
explanatory: <TITLE> text </TITLE> delineates a title, the second tag,
preceded by a slash cancelling the first tag. <HEAD>, </HEAD>, <BODY>,
</BODY> tags act in exactly the same way. <EM> adds emphasis to text.
Browsers differ on their interpretation of this last type of tag, which is
known as a style tag. In this case the text is usually underlined or italicized
or both. The tag <P> instructs the browser to begin a new paragraph. This
type of tag – which requires no subsequent end tag – is known as an
empty tag. Other empty tags include <BR>, which inserts a line break, and
<HR>, which inserts a horizontal rule.

Anchor tags

Anchor tags are the ends of hypertext links. These mark-up tags alone are
the very essence of hypertext. An anchor tag <A...> must have at least
one of two attributes; either a Hypertext REFerence (HREF) or a NAME. If
the destination of a hypertext link is text within the same document,
the NAME attribute is used to distinguish the text as that specified by the
original HREF attribute. For instance, if you use

Click on <A HREF = "multimedia"> MULTIMEDIA </A> to read
more about it

as an anchor in a document, then somewhere else in the same document
you must use a NAME anchor like this:

<A NAME = "multimedia"> MULTIMEDIA </A> </H1>

The first attribute link will cause the word MULTIMEDIA to be specially
highlighted (usually coloured in blue and underlined). Clicking on the first

multimedia link will cause the browser to jump to the named anchor link. Note that when the HREF attribute is a name, it is necessary to precede it with a crosshatch sign. Also note that if the named link is also a heading, or is marked up in some other way, the anchor tag is always written as the innermost tag.

If the destination of a link is another file within the same directory, then all that is needed is an anchor with an HREF attribute set as the file name. A link made this way will read

```
Click on <A HREF = "mmedia.htm"> MULTIMEDIA </A> to read
more about it
```

which will cause the browser to jump to the file MMEDIA.HTM. When a source and destination are in the same directory on a web server, a link made between them is known as a relative link, and the addressing method (like that shown above) is known as relative addressing. Provided all the files stay together in the same logical directory, none of the links need ever be re-specified.

Images

Images represent the first step from a hypertext document to a hypermedia document. An image can be included by means of the <IMG> tag. This is an empty tag, so no text is enclosed and no ending image tag is defined. A typical image tag will read

```
<IMG SRC = "PICTURE.GIF">
```

where SRC is the source attribute and has the same syntax as the HREF attribute discussed above. The image specified is GIF, which is an image in the graphics interchange format (already met in Chapter 8). GIF format is the recommended format for images in an HTML document.

MPEG-IV – object-oriented television coding

Now we come to look at the possible future of television audio and video coding. We have already seen how the incorporation of HTML pages begins to blur the distinction between television and the WWW. The MPEG-IV standard addresses the coded representation of both natural and synthetic (computer-generated) audio and visual objects, and thereby blurs the distinction between CGI, virtual reality and television. MPEG-IV will support all present television and computer video standards, and is optimized for very low bitrates – from 4 Mbits/s down to 64 kbits/s! In fact, at the beginning of the work on MPEG-IV, the objective of the new standard was to address very low bitrate coding issues relating to video conferencing, for example. However, its scope has considerably widened to include even broadcast television applications.

Objects and scenes

Thinking abstractly for a moment, what have MPEG-I and MPEG-II got in common? Where are they 'moving' television? In the direction of a more intelligent receiver and a lower bandwidth transmission link. A DCT representation of a scene is a higher level abstraction of that scene than was the amplitude representation; this requires the television set be much 'smarter' than it was but it does mean the amount of data to be transmitted is very greatly reduced. MPEG-IV (remember MPEG-III disappeared inside MPEG-II) takes this abstraction even further. MPEG-IV is an object-oriented coding standard for television pictures. In terms of functionality, it provides for the shape coding of arbitrarily shaped objects, sprite generation and rendering systems at the decoder. In Chapter 8 we saw the difference between bitmap graphics and vector (or object-oriented) graphics; it is this distinction that separates MPEG-IV from its precursors. MPEG-IV is an evolving standard. Nevertheless, I hope the following will give a taste of where television will be in the future.

MPEG-IV will define a method of describing objects (both visual and audible) and how they are composited and interact together to form 'scenes'. The scene description part of the MPEG-IV standard describes a format for transmitting the spatio-temporal positioning information that describes how individual audio-visual objects are composed within a scene (see Chapter 8 for a revision of the concepts of 3D graphics and the mathematics used). So what does this mean? It means that, for instance, a children's cartoon programme might not be sent as 1500 pictures every minute but in terms of a background description, character descriptions and vectors indicating movements at appropriate times. In other words, the actions of the animator would be sent as high level commands, leaving the MPEG-IV TV set to do the visualization and rendering.

You might be thinking: 'But how would this relate to live television?' A simple example would be the segmentation of the background and foreground object, which could be a rough segmentation performed at real-time. For example, broadcast techniques like chroma-key could be used to gain pre-segmented video material. Other applications of segmentation techniques include object tracking. This simple example, combined with virtual sets technology (see Chapter 8), would offer the intriguing opportunity, for instance, to substitute the news reader you prefer.

The language

What we didn't consider in Chapter 8 was a formalized language for the construction and manipulation of 3D graphics. One such language is virtual reality modelling language (VRML), and the proposed MPEG-IV scene description has several similarities to VRML. For that reason, VRML is briefly described below in order to give a taste of how MPEG-IV will treat 3D video information in terms of objects and scenes. Beware, however, that current MPEG-IV proposals define a far more flexible environment than that described below.

Virtual reality modelling language (VRML)

It is accurate to say that VRML is to virtual reality what HTML is to multimedia development. That is, it is a widely available (and largely free) programming language that permits the construction of virtual 3D worlds, which may be accessed via the Internet/World Wide Web. Just as with the World Wide Web, the creation of such a 3D information hyper-dimensional environment involved two vital components; a language in which to author the environments and the creation of readily available browser programs so that visiting this information space would be more than an experience for a highbrow elite.

VRML browsers

VRML document browser programs come in various types. The simplest is a helper program that is 'launched' by an existing HTML graphical browser when it detects a valid VRML file extension (.WRL), in just the same way the Windows Media Player is launched from the graphical browser when an AVI extension is encountered, provided this has been set up in the browser program configuration. A second type of VRML browser is 'network-aware'; in other words, it has the same communications 'back-end' as does a standard HTML browser. Such a stand-alone application deals with the necessary network protocols. A third type seamlessly integrates VRML browsing with HTML browsing, as well as incorporating other applications in a manner that is transparent to the user. Essentially, an MPEG-IV TV would incorporate a VRML browser as part of its decoder.

Importantly, VRML envisages viewer participation (interaction), as does MPEG-IV. Looking at contemporary WWW type browsers, the manner in which participant navigation is controlled within the 3D environment differs from program to program (each having its own form of control metaphor). Template Graphics Software's WebSpace (a helper application) employs a steering handle analogy and pitch knob; Intervista's WorldView (a stand-alone, network-aware browser) uses fly, translate and tilt buttons. Each metaphor has its advantages and disadvantages, yet all seem easy to use. This is an area of such frantic activity that changes are being implemented constantly. Figure 12.1 is an off-screen shot of Intervista's WorldView.

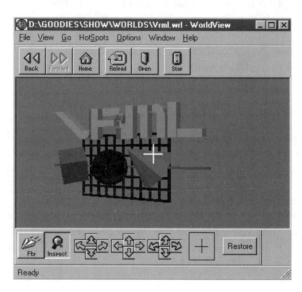

Figure 12.1   *Intervista's WorldView, stand-alone VRML browser*

VRML was conceived in the spring of 1994 at the first annual World Wide Web Conference in Geneva, Switzerland, at which several attendees described projects underway to build three-dimensional graphical visualization tools that inter-operate with the Web. Everyone agreed on the need for these tools to have a common language for specifying 3D scene description and WWW hyperlinks. The term virtual reality mark-up language (VRML) was coined, and the group resolved to begin specification work after the conference. The word 'mark-up' was later changed to 'modelling' to reflect the graphical nature of VRML. From very early on, it was hoped that VRML could be adapted from an existing solution. A set of requirements for the first version was quickly agreed upon, and thus began a search for technologies that could be adapted to fit the needs of VRML. The search turned up several worthwhile candidates, one of which was the Open Inventor ASCII File Format from Silicon Graphics, Inc. The Inventor File Format supports complete descriptions of 3D scenes with polygonally rendered objects, lighting, materials, ambient properties and realism effects. This was the existing solution launch-pad for VRML, which is in effect a subset of the Inventor File Format, with extensions to support networking. Gavin Bell, of Silicon Graphics Inc., adapted the Inventor File Format for VRML. SGI has publicly stated that the file format is available for use in the open market, and contributed a file format parser into the public domain to bootstrap VRML viewer development.

VRML 1.0 is designed to meet the following requirements: platform independence, extendibility and ability to work well over low-bandwidth connections. Every VRML file must be a standard ASCII file. Any text in a file which is preceded with a # is commented out until the next line-return. All the information within the file must be text – that's it; no bitmaps, no special characters – just like an HTML file. In addition, each file must begin with a standard header

```
#VRML V1.0 ascii
```

(Note that a VRML browser will not parse a document without this header, even though the appearance of the # sign seems to imply the header is commented out!)

The 3D environment described within a VRML document is really nothing more than a list of objects, with each object having various attributes. Each object is termed a node, and the attributes are termed fields. Nodes may be embedded. The number of fields a node may have depends on the type of node. The entire list of nodes and fields is known as a scene graph. Scene graphs are hierarchical; in other words, it's not only the nodes themselves but also the order in which they appear that matters in a scene graph. Technically, this is referred to as the notion of state. A scene graph is said to have a notion of state because nodes earlier in the scene may affect nodes that appear later. A mechanism is incorporated within the language to delimit the effect of earlier nodes upon later ones, thus allowing part of the scene graph to be functionally isolated from the other parts. This mechanism is mediated by a special class of nodes called group nodes. Types of simple shape nodes include sphere, cube, cylinder and cone, amongst others. Taking the sphere as the first example; it has only one attribute, or field, that describes radius. The basic syntax for all VRML nodes is

```
objectname objecttype { fields}
```

Object names are often not used; object type and the curly brackets are required. Take a simple example of a 3D scene graph. The following file is happily browsed as a complete 3Da.e. description (although perhaps not a very interesting one!):

```
#VRML V1.0 ascii

Sphere {
radius 2
}
```

Note that one word space (known by printers and software engineers as white-space) separates each of the syntactical entries in a VRML file. Extra white-space is ignored, and software engineers often insert many extra line returns and tabulate their files to make them easier to read. Browsing the tiny file listed above, the computer displays a whitish ball in the centre of a light blue viewing window. Clearly we haven't told the computer very much about our 3D world; the position of the camera, the colour of the sphere etc. This helps introduce a number of important defaults that are part of the VRML specification: A right-handed, three-dimensional co-ordinate system is employed. By default, objects are projected onto a two-dimensional plane in the direction of the positive z axis. The standard unit for physical measurement is metres. Angles are specified in radians. Note that the navigation controls, which are clearly visible in Figure 12.1 for example, navigate the position of the camera in the 3Da.e. (see Chapter 8). The default camera location when a file is loaded, known as the entry point, is at the x, y origin and 1 metre back (out of the screen), looking along the negative z axis – i.e. (0, 0, 1). Furthermore, the fact that the ball is whitish in a blue environment demonstrates that some material defaults are at work. We shall see how many of these parameters may be altered, using simple text-based commands, in order to add diversity to a virtual world.

In all, VRML defines 36 different types of nodes, divided into eight different classes; shape nodes, geometry and material nodes, transformation nodes, camera nodes, lighting nodes, group nodes and a miscellaneous class with one member.

Shape nodes

Shape nodes define both basic shapes (these primitives include cone, sphere, cube, cylinder and text) and much more complex and unusual (real-world) shapes to be defined in terms of arrays of polygons, lines or points. The sphere has already featured as an example. The cube is given as a further example. If a cube is called up like this:

```
Cube {
}
```

it could hardly be more straightforward! The parser defaults to a cuboid aligned with the co-ordinate axis and measures 2 units in each direction. The cube is drawn at the position of the current translation, and rendered with the current material and texture (explained later). Different sizes of cuboid are defined using width, height and depth fields, like this:

```
Cube {

width 8.3
depth 8.254
height 10.05

}
```

Each value is a single-precision floating-point number.

The cone node represents a simple cone whose central axis is aligned with the y axis. By default, the cone is centred at (0, 0, 0). The cone has a base with diameter 2 and height of 2. The cone is drawn at the position of the current translation, and rendered with the current material and texture. A cone comes in 'two parts', the base and the sides. If the conical part alone is required (i.e. looking like a horn), then this is defined using the parts field – like this:

```
Cone {
parts SIDES
bottomRadius 1.05
height 2.78
}
```

If, instead, a conical volume is required (i.e. with a base) the following would be used:

```
Cone {
parts ALL
bottomRadius 1.05
height 2.78
}
```

The cylinder too has the ability to be split into parts; this time into SIDES (the cylindrical 'drainpipe' part), TOP, BOTTOM and ALL. A cylindrical volume would be expressed like this:

```
Cylinder {
parts ALL
radius 2.78
height 5.098
}
```

Note that if various parts are required these are written this way:

```
Cylinder {
parts ( SIDES | BOTTOM )
radius 2.78
height 5.098
}
```

Geometry and material nodes

Geometry and material nodes affect the way shapes are drawn. One of the most important nodes in this class is the Material node. This defines the current surface material properties for all subsequent shapes. An example of the material node at work would be

```
Material {
diffuseColor 0 0 1 # These three figures are R, G, B values.
shininess 0.5
}
```

which would specify a fairly shiny blue colour for all subsequent objects.

Texture mapping

The techniques involved in the rendering of surfaces were covered in Chapter 8, and browsers usually have options for the employment of wireframe, flat shading, Gouraud or Phong shading. However, whilst it would be theoretically possible to define any surface texture in terms of areas of colour, reflectance and so on, this would involve any TV/computer (even the most advanced machines available) in a gargantuan mathematical task. In any case, this computational overhead is not really necessary if a technique known as texture mapping is employed. In texture mapping, a bitmap 'wallpaper' is applied to the polygon faces (each of which appear suitably transformed after projection). Often an entire object is 'wallpapered' with a single texture bitmap. Employed a great deal in all 3D graphics systems, texture mapping is very computationally efficient. VRML is an ideal application, and the language supports texture mapping in the following ways. In the case of the cube, textures are applied individually to each face of the cube; the entire texture goes on each face. When texture is applied to a cylinder, it is applied anticlockwise about the sides with a vertical seam at the back. For the top and bottom, a circle is cut out of the texture square and applied to the top or bottom circle. When a texture is applied to a sphere, the texture covers the whole surface, wrapping anti-clockwise from the back with the vertical seam is at the back of the sphere.

Texture mapping, initiated by the Texture2 node, defines a texture map and parameters for that map and instructs the browser to apply the texture map to all subsequent shapes – unless delimited as explained below. Texture can be read from a logical file name or from a URL (over the Web), and this location is defined in the file-name field.

Transformation nodes

Transformation nodes include MatrixTransform, Rotation, Scale, Transform and Translation. Taking Translation as an example; this node defines a translation by a 3D vector, everything written into the scene graph after such a node is translated by the value of this vector unless a grouping node is included to prevent this.

Camera nodes

Camera nodes are used to define the projection from a viewpoint. Two nodes, OrthographicCamera and PerspectiveCamera, define a projection either where objects do not diminish in size with distance or where they do, respectively. The default is the orthographic projection. This node also provides the ability to define the position of the camera within the 3D co-ordinate system.

Lighting nodes

This set of nodes defines the position and type of light that falls on the 3D scene. Three nodes are provided; DirectionalLight, PointLight and Spot-Light. Definitions of point light sources and spot lights were given in Chapter 8. The DirectionalLight is a third type of light that acts like a laser and thereby illuminates along parallel rays to a given three-dimensional vector. A point light source is defined thus:

```
PointLight {
            on TRUE
            intensity 1
            colour 1 1 1
            location 0 0 1
      }
```

which happens to be the default value.

Group nodes

Remember that scene graphs are hierarchical and that a scene graph is said to have a notion of state because nodes earlier in the scene may affect nodes which appear later. The mechanism, incorporated within the language to delimit the effect of earlier nodes upon later ones – thus allowing part of the scene graph to be functionally isolated from the other parts – is mediated by the group nodes, of which the Separator node is the most important. The other important group node is the WWWAnchor node. This is the heart of synthesis of 3D graphics and the Web, because it is this node that loads a new scene into a VRML browser by defining a URL when a shape is (usually) clicked with a mouse. The WWWAnchor acts, itself, like a Separator node in that it effectively isolates the section of the scene graph it delimits.

Miscellaneous nodes

In HTML, an image within an HTML document can be defined in terms of an URL, like this:

```
<img src="http://www.company.org/picture.gif">
```

so that, when a browser encounters such an image mark-up tag, it goes off and gets the image from the desired Internet location. Such an image is said to be an inline image, and the technique is known as image inlining. A similar technique is possible in VRML, and the WWWInline node provides for this.

Practical VRML files

Knowing a little about the workings of VRML file syntax is worthwhile, and it's very satisfying to do a little typing and have the results of your endeavours appear before your eyes. Pesce's book *VRLM – Browsing and Building Cyberspace* (1995) provides an excellent description of the language. A simple example is given below, along with a view of the scene graph rendered with WorldView (see Figure 12.2). Note particularly the way the texture mapping works, as described above.

```
#VRML V1.0 ascii

Separator {
Material {
emissiveColor 1 1 0
}
Texture2 {
        filename "c:\windows\clouds.bmp"
}
Cube {
height 2
width 2
```

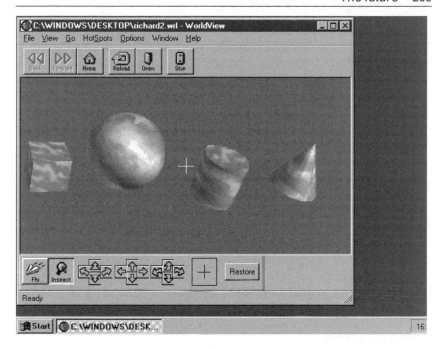

Figure 12.2 *A simple VRML 3DAE rendered in InterVista's WorldView*

```
depth 2
}
Transform {
translation 3 3 3
}

Sphere {
radius 2
}
Transform {
translation 2 2 5
}
Cylinder {
}
Transform {
translation 2 2 2
}
Cone {
}
}
```

However, it's important to make a distinction between HTML and VRML authoring. Originally, HTML documents had to be written using a text-based editing program, and although nowadays many tools exist to assist in the creation of HTML pages, writing 'by hand' still remains a viable authoring technique. Not so with VRML. Interesting (realistic) shapes may require thousands of polygons to be defined – a task quite beyond even the most patient typist! Fortunately there are many 3D graphics applications programs that provide a more user-friendly interface for the construction of 3D worlds. Of course, SGI's Open Inventor was not the only choice for a 3D scene description language. In fact, languages are legion, and very few of them interoperate. Each application tends to have its own native format, and it becomes necessary to convert from one to the other as a project moves from application to application. Increasingly, three-dimensional graphics applications furnish the capability of saving a description of the scene in VRML form. It is a measure of the success of VRML that it has begun to be regarded as a 'metafile data format' (like GIF files have for images) for the universal interchange of 3D data. The very same reason lies behind the choice of VRML as the model for MPEG-IV picture coding.

MPEG-IV audio

MPEG-IV defines audio in terms of objects too. In fact, a 'real world' audio object is defined as an audible semantic entity recorded with one microphone in the case of a mono recording, or with more microphones at different positions in the case of a multi-channel recording. Audio objects can be grouped or mixed together, but objects can not easily be split into sub-objects. Applications for MPEG-IV audio might include 'mix minus 1' applications, in which an orchestra is recorded minus the concerto instrument, allowing viewers to play along with their instruments at home. Or where all effects and music tracks in a feature film are 'mix minus the dialogue', allowing very flexible multilingual applications because each language is a separate audio object and can be selected as required in the decoder.

In principle, none of these applications is anything but straightforward. They could be handled by existing digital (or analogue) systems. The problem, once again, is bandwidth. MPEG-IV is designed for very low bit-rates, and this should suggest that MPEG has designed (or integrated) a number of very powerful audio tools to reduce necessary data throughput. These tools include the MPEG-IV structured audio format, which uses low bit-rate algorithmic sound models to code sounds. Furthermore, MPEG-IV includes the functionality to use and control post-production panning and reverberation effects at the decoder as well as the use of a SAOL signal-processing language enabling music synthesis and sound-effects to be generated, once again at the terminal; rather than prior to transmission.

Structured audio

We have already seen how MPEG (and Dolby) coding aims to remove perceptual redundancy from an audio signal; as well as removing other simpler representational redundancy by means of efficient bit-coding schemes. Structured audio (SA) compression schemes compress sound by, first, exploiting another type of redundancy in signals – structural redundancy.

Structural redundancy is a natural result of the way sound is created in human situations. The same sounds, or sounds that are very similar, occur over and over again. For example, the performance of a work for solo piano consists of many piano notes. Each time the performer strikes the middle C key on the piano, a very similar sound is created by the piano's mechanism. To a first approximation, we could view the sound as exactly the same upon each strike; to a closer one, we could view it as the same except for the velocity with which the key is struck, and so on. In a PCM representation of the piano performance, each note is treated as a completely independent entity; each time the middle C is struck, the sound of that note is independently represented in the data sequence. This is even true in a perceptual coding of the sound. The representation has been compressed, but the structural redundancy present in re-representing the same note as different events has not been removed.

In structured coding, we assume that each occurrence of a particular note is the same, except for a difference that is described by an algorithm with a few parameters. In the model-transmission stage, we transmit the basic sound (either a sound sample or another algorithm) and the algorithm that describes the differences. Then, for sound transmission, we need only code the note desired, the time of occurrence, and the parameters controlling the differentiating algorithm (Scheirer, 1997).

Structured audio orchestra language

SAOL (pronounced 'sail') stands for structured audio orchestra language, and it falls into the music-synthesis category of 'Music V' languages. Its fundamental processing model is based on the interaction of oscillators running at various rates. Note that this approach is different from the idea (used in the multimedia world) of using MIDI information to drive synthesis chips on soundcards. This latter approach has the disadvantage that, depending on IC technology, music will sound different depending on which soundcard is realized. Using SAOL (a much 'lower-level' language than MIDI), realizations will always sound the same.

At the beginning of an MPEG-IV session involving SA, the server transmits to the client a stream information header, which contains a number of data elements. The most important of these is the orchestra chunk, which contains a tokenized representation of a program written in structured audio orchestra language. The orchestra chunk consists of the description of a number of instruments. Each instrument is a single parametric signal-processing element that maps a set of parametric controls to a sound. For example, a SAOL instrument might describe a physical model of a plucked string. The model is transmitted through code, which implements it, using the repertoire of delay lines, digital filters, fractional-delay interpolators and so forth that are the basic building blocks of SAOL.

The bitstream data itself, which follows the header, is made up mainly of time-stamped parametric events. Each event refers to an instrument described in the orchestra chunk in the header, and provides the parameters required for that instrument. Other sorts of data may also be conveyed in the bitstream; tempo and pitch changes, for example.

Unfortunately, as at the time of writing (and probably for some time beyond!), the techniques required for automatically producing a structured audio bitstream from an arbitrary pre-recorded sound are beyond today's state of the art, although they are an active research topic. These techniques are often called 'automatic source separation' or 'automatic transcription'. In the meantime, composers and sound designers will use special content creation tools to directly create structured audio bitstreams. This is not considered to be a fundamental obstacle to the use of MPEG-IV structured audio, because these tools are very similar to the ones that contemporary composers and editors use already; all that is required is to make their tools capable of producing MPEG-IV output bitstreams. There is an interesting corollary here with MPEG-IV for video for, whilst we are not yet capable of integrating and coding real-world images and sounds, there are immediate applications for directly synthesized programmes.

Text-to-speech systems

MPEG-IV audio also foresees the use of text-to-speech (TTS) conversion systems. There are two commonly used techniques for generating speech. The first is table-based, and works like a dictionary. Each word is stored in both a text rendering and a sound rendering. A lookup is performed on the text version, and the corresponding sound version is selected and played back. This is really a 'belt-and-braces' technique. The second technique is more subtle. No text is stored and little or no voice recording is necessary; just the rules used to convert from text to speech. Such a technique is called rule-based. The rules are used to convert text to a set of sound descriptors (phonemes) which, when played via a loudspeaker, are heard as speech. A rule-based program can always pronounce any word it encounters. However, because English is an irregular language, words that don't follow the rules will be mispronounced. For this reason, practical TTS systems often have to revert to a table-based system for exceptions. Table 12.1 is a list of the complete phonemes used in an English TTS system.

Table 12.1 *List of the complete phonemes used in an English TTS system*

Phoneme	Pronunciation
IY	as in beet
IH	as in bit
IX	as in decide
EH	as in bet
AE	as in bat
AX	as in about
AA	as in cot
UH	as in book
UW	as in boot
OW	as in boat
ER	as in bird
AY	as in bite
EY	as in bait
OY	as in boy
AW	as in bout
LX	as in tall
l	as in low
m	as in mow
n	as in no
NG	as in sing
y	as in yes
r	as in red
w	as in wed
b	as in bed
d	as in dead
g	as in get
v	as in vet
DH	as in then
z	as in zen
ZH	as in usual
f	as in fit
TH	as in thin
s	as in sin

SH	as in shin
h	as in him
p	as in pin
PX	as in spin
t	as in top
TX	as in stop
DX	as in butter
k	as in kite
KX	as in sky

In addition, programs provide a number of phonetic modifiers which involve instructions to shorten or lengthen individual phonemes, to increase or decrease pitch etc. The action of these modifiers is to modulate the speech (an action known as prosody), and they greatly improve the naturalness of the speech produced. Individual voices are very often associated with particular inflections and prosody. The MPEG-IV TTS can not only synthesize speech according to the input speech with a rule-generated prosody, like the simple system described above, but also executes several other functions including:

• Speech synthesis with the original prosody (this being extracted from the original audio input in the process of coding)
• Synchronized speech synthesis with facial animation (FA) tools, which will be used to provide lip-movement information so that, for example, an American film dubbed into French will appear with the appropriate lip movements
• Trick mode functions such as stop, resume, forward, backward without breaking the prosody, even in the applications with facial animation (FA)
• The ability to change the replaying speed, tone, volume and the speaker's sex, and age.

The MPEG-IV TTS has provided the capacity to suit all languages. MPEG-IV speech coders operate at bit-rates between 2–24 kb/s for the 8 kHz bandwidth mode and 14–24 kb/s for the 16 kHz bandwidth mode. This is obviously an enormous coding advantage, even over the most efficient audio coders, and alone justifies their inclusion in the standard.

Audio scenes

Just as video scenes are made from visual objects, audio scenes may be usefully described as the spatio-temporal combination of audio objects. An 'audio object' is a single audio stream coded using one of the MPEG-IV coding tools, such as structured audio. Audio objects are related to each other by mixing, effects processing, switching and delaying them, and may be panned to a particular 3D location. The effects processing is described abstractly in terms of a signal-processing language – the same language used for structured audio.

MPEG-VII and metadata

The global information explosion has raised an important question; what's the point of having access to all the information on the planet if you don't know where it is? The Internet and World Wide Web (WWW) provide an excellent case in point. Most information on the WWW is text, and it would be quite impossible to find required information were it not for text-based search engines like AltaVista and Yahoo. Indeed, text-based search-engine sites are among the most visited sites on the Internet.

With the increasing digitization of television we can expect, as the next century progresses, a gradual explosion of television-type (audio-visual) information in the same way as we have seen with text and still pictures in the last 10 years. But how will the video equivalent of AltaVista work? No generally recognized description of television-type material exists. In general, it is not possible to search for 'Rick, Elsa and aeroplane'! Similarly, at the moment, you can't enter, 'Eb, Eb, Eb, C' and have the search-engine suggest 'Beethoven, Symphony 5 in C minor'!

In October 1996, MPEG started a project to provide a solution to the questions described above. The new member of the MPEG family, called multimedia content description interface (MPEG-VII), will specify a standard set of descriptors that can be used to describe various types of multimedia information. AV material that has MPEG-VII data associated with it will thereby be indexed and could be searched for. This material may include still pictures, graphics, 3D models, audio, speech, video and information about how these elements are combined in a multimedia presentation (known as 'scenarios'). Special cases of these general data types may include facial expressions and personal characteristics. MPEG-VII descriptors do not, however, depend on the ways the described content is coded or stored. It is possible to attach an MPEG-VII description to an analogue movie or to a picture that is printed on paper. MPEG-VII information is therefore a sub-set of metadata, which is defined as 'data about data'.

MPEG-VII builds on MPEG-IV, which provides the means to encode audio-visual material (as we have seen) as objects having certain relations in time and space. Using MPEG-IV encoding, it will be possible to attach descriptions to elements (objects) within the scene, such as audio and visual objects. Indeed, it's possible to see MPEG-VII's place in the family in its connections with MPEG-I and II, for what is block matching and vector assignment but a relatively high-level abstraction of a moving scene? Audio descriptions might be in the form of tempo, mood, key, tempo change etc. The exact syntax of MPEG-VII, as well as the degrees of abstraction that might be allowed, are still the subject of debate at the time of writing.

References

Pesce, M. (1995). *VRML – Browsing and Building Cyberspace*. New Riders Publishing.

Scheirer, E. (1997). eds@media.mit.ed

Index